DISCLAIMER

The information provided in this book should not be construed as personal medical advice or instruction. It is intended for educational purposes only and is not meant to diagnose, treat or prevent any disease. No action should be taken based solely on the contents of this book. The referenced information is obtained from sources believed to be reliable. Licensed health care professionals should rely on sound professional judgment when recommending herbs and natural medicines to specific individuals. Readers should consult appropriate health professionals on any matter relating to their health and well-being, including individual use of herbs and natural medicines. The use of any specific product should always be in accordance with the manufacturer's directions.

SEX, LIES & CHOLESTEROL

DR. RYAN E. BENTLEY

AuthorHouse™
1663 Liberty Drive
Bloomington, IN 47403
www.authorhouse.com
Phone: 1-800-839-8640

© *2010 Dr. Ryan E. Bentley. All rights reserved.*

No part of this book may be reproduced, stored in a retrieval system, or transmitted by any means without the written permission of the author.

First published by AuthorHouse 9/23/2010

ISBN: 978-1-4520-7222-7 (e)
ISBN: 978-1-4520-7221-0 (sc)
ISBN: 978-1-4520-7220-3 (hc)

Library of Congress Control Number: 2010912390

Printed in the United States of America

This book is printed on acid-free paper.

Because of the dynamic nature of the Internet, any Web addresses or links contained in this book may have changed since publication and may no longer be valid. The views expressed in this work are solely those of the author and do not necessarily reflect the views of the publisher, and the publisher hereby disclaims any responsibility for them.

CONTENTS

Disclaimer i
Introduction vii
Chapter 1. The Integrated Body 1
Chapter 2. Lies about Cholesterol 11
Chapter 3. What Drives Cholesterol Production? 21
Chapter 4. The Problem with Statin Drugs 27
Chapter 5. The Wolves Guarding the Chicken Coupe 39
Chapter 6. Deficiencies caused by Cholesterol Lowering Drugs 51
Chapter 7. Lifestyle and Disease 61
Chapter 8. Mapping out a Course to Better Health 69
Chapter 9. Regenerating your Body through Lifestyle Changes 79
Chapter 10. Supplements that Enhance Lifestyle Changes 91
Chapter 11. Summary and Conclusion 107
Reference List 111
Appendix 117

INTRODUCTION

Be honest. Were you inspired to read this book because it has 'sex' in the title? Sorry to disappoint, but this is not a book about sexuality. We *will* deal with sex hormones, however, and the effect that cholesterol has on them. I intend to show that cholesterol is absolutely necessary in many different body functions, including sex hormones and sexual function, and that we lower cholesterol to our own hurt.

I know that's a radical statement, but I also intend to back up that assertion with research. Much of what you have been led to believe about cholesterol is absolutely false, as I will demonstrate, and I hope that what you get out of this book is to be more aware of the dangers of cholesterol drugs and that you should *not* be terrified of cholesterol, as cholesterol is actually beneficial.

Before going any further, let's transition here and talk about something I think you can relate to: the perspective of children.

My children tend to be the compassionate sort. My oldest son has a lot of love in his heart; he just really loves people and genuinely cares for them. He never met his great-grandfather, but he always asks about him because my wife has mentioned him many times. One day he asked "Dad, what happened to Great Grandpa?"

(Keep in mind he is a very analytical child, so I give him every detail that I can possibly think of when he asks a question.)

I said, "Well, Great Grandpa died of a heart attack."

He asked, "What's a heart attack?"

"It's when you don't get blood and oxygen to your heart because something blocks it," I replied.

He then asked, "Why does that happen?"

"Well, some people have certain genes, and when you put them in a certain environment those genes are triggered and it can increase the risk of having a heart attack."

And he responded simply, "Oh." After a pause he added, "So if you have bad genes, you are going to have a heart attack?"

At this point, more explanation was needed. I said, "Well, just because you have a set of bad genes does not necessarily mean that you are going to die of a heart attack; it depends on the environment you put them in. One example might be eating a lot of junk food…"

We play a game at home called, *Man Food versus*

God Food, to determine which type of food leads to good health. As I continued to talk he just sat there nodding his head, and I'm thinking, "Man, my kid is smart; he's getting this; he understands genetics and the environment [epigenetics], and that there is genetic outcome related to the quality of food we consume."

After awhile he looked at me and said, "Dad?"

"Yes," I said, anticipating another insightful remark from my brilliant son.

"What was wrong with Great Grandpa's pants?"

Chuckling, I said, "No, his *genes*, not his jeans."

Don't you just love the perspective of an innocent, learning mind!

It would be great if we could push the re-set button in our minds and just stand back and look at this whole topic from a fresh perspective, like a little child. But since that's not really possible, I will ask you to keep an open mind as we proceed.

In my private practice where I work with all different kinds of patients ranging from the chronically ill to professional and Olympic athletes, I discovered that there is a misconception about cholesterol and its role in cardiovascular disease and overall health, and thus I began to re-educate my patients on this important issue. That re-education inspired me to develop a lecture to physicians on the topic, and eventually I converted the lecture into the book you now hold in your hands.

In this book, we are going to be talking about the misconceptions and sometimes outright lies involved in what has been appropriately called "the cholesterol myth" by some authors. As you learn how all this fits together in the world of modern medicine, I hope to lead you to an in-depth understanding of the true cause of heart disease and why it is the number one killer in America right now. It's a big issue in this country at this time. Heart disease has been the number one killer for twenty plus years and traditional western medicine has made little progress to date to actually fix or prevent the cause of it.

Orville Wright once said something very profound that we can apply to this discussion. *"If we all worked on the assumption that what is accepted as true is really true, there would be little hope of advance."*

So let's look all the way back to circa 1900 for something accepted as true during that era. Heroin was then deemed a non-addictive substitute for morphine and was used as a cough suppressant for kids! Is that still considered true today? Of course it isn't. Doctors don't use heroin in hospitals anymore, because we know now that heroin is indeed an extremely addictive and hallucinogenic substance. So we have dumped it and reverted back to using morphine in hospitals.

So we need to keep that in mind as we ponder what is "true." What we learn today via marketing is not necessarily the real truth; it's marketing. And the primary objective of marketing is profit. Even what is considered the cutting edge of science today is most likely going to be antiquated tomorrow and pitched to the curb. It's always been that way. We learn and progress and advance, hopefully. In doing so we discard old ideas and practices that are proven wrong.

As it pertains to the specific subject of this book, you are going to learn as you read this material that something similar is happening today with cholesterol-lowering statin drugs and many other drugs that are used as "treatment" for lifestyle diseases. As the truth is revealed, I believe we're going to eventually look back laughing and say, "A statin drug for heart disease? Are you kidding me? That is nonsense!"

Bobby Kennedy once said *"Few men are willing to brave the disapproval of their fellows, the censure of their colleagues, the wrath of their society. Moral courage is a rarer commodity than bravery in battle or great intelligence. Yet it is the one essential, vital quality of those who seek to change the world which yields most painfully to change."*

In other words, many people are resistant to change and progress because embracing that change would mean disapproval from their friends, family, colleagues, and/or society. It takes the bravest souls to stand up in a time of great moral weakness and declare a truth in the face of fierce opposition. But it is these same brave souls who have paved the road of progress down through history that has enabled us to enjoy many of the benefits of our modern age.

A vivid example of this is the story of Hungarian physician, Ignaz Semmelweis. In 1844, Semmelweis was appointed assistant lecturer in the First Obstetric Division of the Vienna General Hospital (Allgemeines Krankenhaus), where medical students received their training. Semmelweis noticed his ward's 16% death rate from fever was substantially higher than the 2% in the Second Division, where midwifery students were trained. Semmelweis also noticed that puerperal fever was rare in women who gave birth before arriving at the hospital. Semmelweis noted that doctors in First Division performed autopsies each morning on women who had died the previous day, but the midwives were not required or allowed to perform such autopsies. He made the connection between the autopsies and puerperal fever after a colleague died of an infection after accidentally cutting his hand while performing an autopsy.

Semmelweis began experimenting with various cleansing agents and, from May 1847, ordered that all doctors and students working in the First Division wash their hands in chlorinated lime solution before starting ward work, and later before each vaginal examination. The death rate from puerperal fever in the division plummeted from 18% in May 1847 to less than 3% in June through November of that same year. While his results were extraordinary, his colleagues responded to him with skepticism and ridicule. He did the same work in St. Rochus hospital in Pest, Hungary and published his findings in 1860, but his discovery was again ignored.

The idea of tiny organisms (germs) invisible to the naked eye was seen as witchery in those days. Today, of course, the existence of germs is common knowledge and globally accepted as true.

So when a new idea is first introduced it is often attacked as ridiculous and outrageous, or ignored altogether. That is the condition we now find ourselves in when it comes to cholesterol's involvement in heart disease. Ironically, it is often the most venomous opponents of those same new ideas that give themselves credit for "discovering" them when they are eventually embraced. I predict that someday the concept of cholesterol as the cause of heart disease will be discarded for the truth, and the very ones who so violently opposed the truths I will share in this book will be many of the same ones who will take credit for the "new discoveries."

I hope you take the information you are about to learn and implement it into your life and help change the lives of some of the people around you, because that is what knowledge should be about; it should be about taking that knowledge and turning it into action and implementation.

CHAPTER 1

THE INTEGRATED BODY

In order to better understand where we are ultimately going in this discussion, we must first lay some groundwork. In the light of present knowledge, it is apparent that disease starts primarily at the atomic level – the smallest and most basic elements of life. To begin to understand this, let's talk about the basics of food.

When our bodies digest food, the primary function of the digestion process is to turn that food into energy. Our bodies break food down into smaller particles such as amino acids, fatty acids, sugars, vitamins, minerals and electrons. So when we eat healthy food – *live* food – like an apple or a carrot, the digestive system breaks it down into smaller molecules that nourish our bodies. These nutrients extracted from food are essential to build and fuel our bodies.

On the other hand, when we eat dead food like a candy bar, we are eating food that is completely processed. This type of "food" is mostly man-made, meaning that it likely contains nothing nourishing that comes from the earth. It's synthetic, made up of substances which do not appear in nature and are therefore not nourishing. They provide calories, yes, which the body can burn for energy, but they provide little nutrition, which is what the body needs to optimally fuel energy metabolism. Thus, those calories are what we refer to as "empty calories."

So immediately when we eat those dead foods, we are robbing our bodies of energy because we are taking energy from our bodies to break them down; and as our bodies are trying to metabolize that junk food, we are not only using extra energy to metabolize that sort of food, but we are also leaching our stored nutrients in order to provide the fuel for the extra energy the body has to use. As this occurs, we thus throw ourselves into deprivation (nutrient and energy debt) because with that type of food we are not replenishing our bodies. Therefore, at the atomic level, when we eat junk food we are immediately causing degeneration and putting our bodies into a catabolic (the process of the breakdown of body tissues) state.

Significant and noticeable degeneration doesn't happen the minute that we consume that junk food – it's not going to be felt immediately. That's because the human body has an amazing ability to adapt. However, perhaps ten or twenty years from now when the body runs out of reserves, we see the effects, sometimes in devastating consequences.

The problem is that our society is inundated with junk foods laden with chemicals, such as the flavor enhancer MSG and artificial sweeteners like aspartame and sucralose, to name just a few, which are neurotoxins, confusing the body and causing it to crave more and more of the junk food, while the individual doesn't even realize what is happening and why they eventually begin to feel so bad.

Let me give you an analogy here for this junk food syndrome. It's much like if you take a frog and place it in a pot of boiling water; it's going to jump right out. On the other hand, if you put the frog in a pot and turn up the heat very slowly, it is somehow unable to notice the subtle changes in temperature, and it will sit in there until it boils to death. And that's where we are right now as the human race; we have become numb to our toxic environment and we are slowly killing ourselves.

So why am I going into all this? What does this have to do with sex, lies and cholesterol? Well, it's got *everything* to do with it. So what is cardiovascular disease? Is it an acute crisis situation? No, it's a chronic disease—a chronic disease that leads to an acute emergency situation. And diet is the major player in whether or not people develop cardiovascular disease. Certainly there are genetic factors at work, and stress is a big piece of the puzzle as well. But if we Americans could change just one thing that would make a major impact on the prevalence of cardiovascular disease, it would be our diets. Yet that is not being preached enough by cardiologists and most of the medical profession. They give the concept of lifestyle modification lip-service for the most part, but are quick to prescribe a drug or resort to a surgical intervention of some sort.

In 2009, *The Archives of Internal Medicine* said, "Nutritional efforts should remain the **principal focus** of chronic disease prevention…" (emphasis added). But how often do cardiovascular disease treatments focus primarily on nutrition? How many times are people hearing that? Seldom, if any at all! That's why I hope that you have the moral courage to step out of the box and tell your friends and family about the truth you are about to read, because I guarantee they are not going to hear this message on TV in the near future.

The Book of Life

In attempting to understand how the body is organized, let's discuss some basics by comparing how the body is organized with how a book is organized. In a book we have letters, which are like the smallest elements at the subatomic level (atoms and electrons). The letters form words just like the subatomic particles form cells. The words form sentences just as the cells form tissues, and those sentences form paragraphs just as tissues form the organs of the body. Then the paragraphs are organized into chapters, much like the organs of the body form organ systems (cardiovascular system, neurological system, etc). And finally, the chapters are organized into a flow of continual thought to form the complete story of the book, just as the organ systems organize in such a way as to form the entire organism – the human body. You might recognize how this organization of life is like a tier in the way it builds up from single cells to a complicated human organism that is totally integrated in its functions.

A major problem in our current health care system is that we have a tendency to focus on one symptom, one chemical reaction, or one hormone, make some assumptions about them, and not take a step back and look at the organism as a whole. It's much like if we only look at the letters (atoms) in a book. Does an individual letter or even an analysis of all the "a's" and "b's" in the print tell us anything about the story? No, it doesn't give you any idea. What about the words (cells)? Do they tell you the story? No, because they are just random words. You may understand a word and what it means, but if you don't know the context in which it is used, that word doesn't mean much in the grander scheme.

Would you understand a book only by individual sentences (tissues)? No.

The paragraphs are equivalent to our organs - they give us more of an idea about the story, but just understanding one organ does not enable you to make sense of the whole organism, just as reading a few paragraphs won't tell you the whole story available in the book.

Remember that we considered that the paragraphs of a book are separated into chapters, which are equal to systems of the body—nervous system, circulatory system and so on. Just as the chapters make up the whole book, the systems make up the whole organism.

Let me give you an example of how things can progress when you look at individual systems rather than the whole organism.

I had a patient who had cerebral palsy and had undergone multiple hip and ankle surgeries. She had a tracheotomy, a stomach tube for feeding and had to be catheterized to go to the restroom. You name it, she had it. When she came in to see me her skin color was blue! Her mother said, "I don't know what to do."

The whole time she was in my office I was thinking that she was in an immediate crisis situation and needed to be receiving emergency care. But her mother said that the cardiologist told them her daughter's problem was pulmonary in nature, but the pulmonologist said that it was a cardiovascular problem. She went on to say that neither of the doctors was willing to communicate with the other because each believed the problem was not their specialty.

So I said, "Well, that's interesting because the last time I looked in a textbook the cardiovascular and pulmonary system worked together."

Needless to say, you can't separate the cardiovascular and pulmonary systems into single specialties because it's just like reading one paragraph of a book and thinking you know the whole story.

We need to take a look at the whole book, but also have an understanding about the letters, words, and the flow that makes up that book because without understanding the definition of a word, you're not going to get an idea of what the book is trying to tell you. Unfortunately, at the patient's expense many doctors today are trained to know the definitions but forget to take an all-encompassing look at the whole body.

Sometimes with doctors it is like they are looking at a fossil and not realizing that they are standing in a dinosaur footprint. Doctors can become so focused on one issue that they miss the bigger picture.

The Importance of Cholesterol

So that leads us to cholesterol. Why do we need cholesterol anyway?

Cholesterol by itself is a soft waxy substance, found among fats in the body. It is integrated into every tissue and is a component of every cell membrane.[1] Humans have approximately one hundred trillion cells. So if cholesterol is integrated into every cell, how important would you consider that to be? Obviously, something which is that integral to every cell has to be important.

What does cholesterol do? It is a precursor to the sex hormones DHEA, testosterone, estrogen, and progesterone.[2] It is also a precursor to our stress hormones such as cortisol.[3]

When under stress the body goes into a "fight or flight" mode for survival. For instance, if you are about to be mugged in a dark alley, you need to be ready to bolt and run or else muster the strength to face the attacker and fight. Stress hormones give us the strength for either choice, but we need cholesterol in order to produce those chemicals, and if we don't have cholesterol we can no longer withstand the various stressors of life because our survival mechanisms are affected.

Here's another example of the importance of cholesterol.

Calcitriol is an active form of vitamin D3.[4] Vitamin D is becoming known as a super-nutrient because of all the biological activities in which it is involved. Our bodies absolutely must be able to metabolize vitamin D properly in order to maintain health. And do you know where vitamin D originates? That's right; it comes from cholesterol.[5] Vitamin D deficiency is a big issue in the scientific community right now because of how it relates to so many different diseases. If vitamin D is made from cholesterol, and a person takes a statin drug to lower cholesterol, what do you think would happen to vitamin D production? I would find it very interesting to know how many people who are taking statin drugs to lower cholesterol are vitamin D deficient.

Cholesterol is also integrated into the part of the nervous system known as the myelin sheaths,[6] which help the brain communicate with the rest of the body. To demonstrate the possible devastating consequences of lowering cholesterol and neurological effects, let's consider Multiple Sclerosis (MS) as an example. MS is an autoimmune disease that is characterized by the deterioration of myelin sheaths in the central nervous system (brain and spinal cord).

To better understand MS, let's take a closer look at the myelin sheaths and their function. If you cut open an extension cord you can see that there is a copper wire inside and a plastic coating around it. That plastic coating is insulation and covers the copper wire. If there were two such wires and they didn't have insulation around them and you let the wires touch, what would happen? They would short circuit, right? Sparks would definitely be flying around.

Essentially the same thing happens with MS. One of the primary clinical diagnostic factors in MS is random pains and sensations in different areas that don't follow a certain pattern. Why is that? Because the neurons are short-circuiting because their myelin sheaths are being eaten away due to autoimmune disease kicking in and throwing the immune system into overdrive.

Research indicates that autoimmune disease of various sorts may be related to Vitamin D deficiency. Additionally, according to WebMD health news, "Vitamin D supplementation, shown in recent studies to help lower risk of certain *cancers, multiple sclerosis, arthritis* and other conditions, may also relieve *depression,* according to new research." This information begs the question that if vitamin D is so important to neurological and immune function, is it really such a good idea to lower the very substance that vitamin D and the myelin sheaths are made from, which is cholesterol?

Is it starting to sink in now? We have, in fact, been told lies about cholesterol, as the truth has been withheld about the many benefits that cholesterol has to the human body. But why in the world would the truth be withheld about such an important health issue? We'll discuss the answer to that question as we proceed throughout the rest of the book.

CHAPTER 2

LIES ABOUT CHOLESTEROL

Before we continue, let me ask you a question so that we are on the same page, so to speak. *Is cholesterol good or bad?* I trust that you can see by now that it is good, right? Indeed, our bodies depend upon cholesterol for many different functions. However, in general most Americans believe cholesterol is bad because that's what marketing and the mainstream media has led us to believe. But remember, marketing is *designed to sell;* and the media thrives on hype and bad news, because that's what sells. In the case of cholesterol-lowering drugs, millions of people have been sold that the drugs are necessary when, in fact, they likely are doing more harm than good.

Understanding Lipoproteins

Before we discuss the specific lies regarding cholesterol and cholesterol-lowering drugs, let's first define some terms and attempt to better understand the role of cholesterol.

Low-density lipoprotein (LDL) is known as the "bad cholesterol." (Actually, the hype regarding cholesterol in the media and throughout conventional medicine is that it's *all* bad if the total is too high, but LDL is supposedly especially nasty!) LDL is simply a protein that has fats attached to it. But here's the question: Is it considered "bad cholesterol" because it is made up of 25% protein, 5% triglycerides, 20% phospholipids, and **50% cholesterol**?[7] The fact that *LDL* is made up of 50% cholesterol is only one of the two reasons that it is known as bad cholesterol.

Consider carefully the purpose of LDL. It carries out 75% of the total cholesterol in the blood and delivers it to the cells for cellular repair and production of hormones. That's right. LDL takes cholesterol to the *one hundred trillion cells* in our body to help the cells repair damage. LDL also carries cholesterol to the organs in the body that make sex hormones (i.e. the testes in men and the ovaries in women). And cholesterol is also very important in the production of bile salts, which help prevent gallstones.

The other lipoprotein, known as high-density lipoprotein (HDL), is known as the "good cholesterol" because it is made up of 45% protein, 30% phospholipids, and **only 20% percent cholesterol**.[8] Also, HDL is considered "good" because it actually eliminates excess cholesterol from the cells and transports it back to the liver for excretion.

To clarify further, let's consider the analogy of dump trucks and bulldozers. LDLs are like the dump trucks that are dropping off the cholesterol, and HDLs are the bulldozers that come in and remove all the excess cholesterol. That is another reason why HDL's and LDL's are considered 'good' and 'bad.' *Actually they are both good and our bodies need them.* Yes, *both* of them! Our cells are repaired with the help of cholesterol, and we make sex hormones and steroid hormones with cholesterol. We also need Vitamin D which comes from cholesterol. We need good bile flow, which occurs with cholesterol, thus preventing gallstones. *We need both HDL's and LDL's equally.*

There are two ways our body gets cholesterol. One way is that the body manufactures it, and the other way is the body absorbs it from the diet.

The HMG CoA Reductase enzyme in the liver makes the majority – 75% of the cholesterol. The other 25% comes from our dietary intake. Many people believe that we shouldn't eat eggs because they are high in cholesterol. Dietary sources do comprise 25% of the total cholesterol, but personally and professionally, I say go ahead and eat eggs because they are an excellent source of protein, and most of the time when a person has high cholesterol it is *not* because of too much dietary cholesterol; it is that the body is making more to compensate for other issues that we will get into shortly.

The Cholesterol Lies

There are some big lies being told about cholesterol and now it's time to talk about them.

Back in the 1980's the concept of increased dietary fat consumption being the cause of elevated cholesterol was popularized. What followed was the low-fat/no-fat diet craze. Keep in mind that every physiology textbook states that the human body is made to run off water, vitamins, minerals, *fats*, proteins, and carbohydrates. In my entire time as a student, professor, and a doctor, I have never seen any textbook that stated otherwise. So let's look at what happens when fat and cholesterol are removed from the body.

When cholesterol is dramatically reduced in the diet, the amount of cholesterol in the blood actually increases! I know that's completely contrary to what you have been led to believe, so I'll try to unravel that mystery for you.

In one study on cholesterol, volunteers were fed 3 to 4 grams of cholesterol daily, which is far more than one could ever obtain from food. As a result of the increase in dietary cholesterol consumption, *the production of cholesterol by the liver was almost completely suppressed!*[9] Basically, when cholesterol was added to the diet in large amounts the body recognized that it had to compensate by lowering its own production in order to keep the cholesterol levels more optimal for biological function.

Similarly, when cholesterol is *removed* from the diet, the body recognizes a drop in the levels and compensates by *increasing* cholesterol production.

To summarize, when cholesterol is low the body makes more, and when cholesterol levels are high a **healthy body** will bring cholesterol levels down. Thus, in my opinion, the low-fat/no-fat diets provided the platform which led to the promotion of statin drugs in the 1980's. On this type of diet, people started to have higher cholesterol levels in their blood even as they vigorously tried to avoid dietary cholesterol and fats!

We have amazing bodies that compensate and adjust according to surrounding internal and external environmental stressors. It's the same thing when we give someone thyroid hormones when they still have a functioning thyroid: the thyroid stops producing its own form of the hormone because it does not need to waste energy making what the body is already getting from an outside source. It's a feedback loop. The body communicates what it needs and what it has in surplus throughout its systems. The human body is very, very smart, and is designed by God to be self-regulating.

Here is another lie: *LDL cholesterol causes heart disease or atherosclerosis.*

The truth is that a degenerative process called, *oxidation*, ultimately leads to an inflammatory process, which can then lead to arterial plaque formation. The fact is that **the majority of people who are dying of heart disease have normal LDL cholesterol levels!** The cause of heart disease is *not* the presence of LDL cholesterol, but that the LDL in the arteries is being oxidized and inflamed year after year, leading to accumulation of plaque in the arteries and an eventual heart attack.

Atherosclerosis is clearly *an inflammatory disease* and not the result of a simple accumulation of lipids.[10] What is the initiator of inflammation? Again, it's oxidation, or free radical damage.

Oxidation is defined as the interaction between oxygen molecules and all the various substances they may contact, from metal to living tissue. Oxidation is what causes metal to rust and rubber to become hard and brittle. It is also what causes the human body to age. While oxidation is a very normal process, there are often disturbances in the natural oxidation process, such as the attraction of a free radical (a renegade molecule) to another molecule in the body, resulting in damage to cells.

Free radicals, also known as reactive oxygen species (ROS), cause oxidation and are molecules that lead to aging, tissue damage, and various degenerative diseases. These free radicals are very unstable, and in attempting to stabilize, they look to bond with other molecules, thus destroying the other molecule's vigor and perpetuating the degenerative process. It can be compared to a drowning person grabbing onto another swimmer in an attempt to stay afloat, but the result is that both of them go under.

Just as oxidation can cause metal to rust, this same degenerative process involving free radicals is, in effect, why the human body "rusts" throughout its lifetime, because free oxygen radicals are running wild through the system seeking to attach themselves to stabilizing molecules.

An example of oxidation with which you are familiar is when you leave a half-eaten apple sitting out and it turns brown. That browning is actually the first stages of the rotting process, and it is due to oxidation/free radical damage.

When cholesterol oxidizes it gets sticky and clumps together, sticking to the arteries. The obvious truth is that it is not cholesterol in and of itself that is bad; it is *oxidized* cholesterol that becomes the issue.

Let me put it this way: A large forest of trees does not guarantee a forest fire, right? Of course it doesn't. However, what if there is a spark that starts a fire in a *small* forest of trees? At this point we have a higher potential for a small forest fire. What if a spark ignites in thousands of acres of trees? There would now be a higher potential for a bigger forest fire. In the same vein, just because a person has high cholesterol, it does not mean that person is going to have a heart attack. If there is a spark – oxidation – there is an increased potential for heart disease.

So is there increased risk involved by having high cholesterol? Potentially, yes. But only because there are more trees for the spark to ignite. Cholesterol can be likened to the trees. More cholesterol can increase the risk to a heart attack only if there is a spark. The spark is not the cholesterol. The spark is the oxidation that leads to the inflammatory process, and that can potentially lead to the forest fire – a heart attack. There is an increased risk of a bigger forest fire (heart attack) because there is more cholesterol circulating around, but the cholesterol itself is not the cause of heart disease. Conversely, people with low or normal levels of cholesterol but who also have a lot of oxidation (the sparks that start the fire) are still at risk of heart disease.

The number of people taking cholesterol lowering statin drugs today is much higher than it has ever been, as we will examine more closely later on. But in 2003 it was stated by the Journal of Cardiology that while over 13 million Americans take drugs to control their cholesterol, **50% of them have normal cholesterol levels and still have heart attacks.**[11] Since this is true, we are forced to consider whether cholesterol-lowering statins actually fix the problem of cardiovascular disease risk. In my opinion the answer is that they most certainly do not.

Spaniards actually have LDL levels similar to Americans but have half the rate of heart disease. The Swiss have higher cholesterol levels, but their heart disease rates are lower than those in the United States.

And Aborigines throughout the world have lower cholesterol levels but higher rates of heart disease.[12] Thus, it would appear that cholesterol numbers truly do not make much difference.

The Role of Genetics

Often I will hear people say that heart disease is genetic. "It runs in my family." The truth is that familial hypercholesterolemia (genetically high cholesterol) at best affects only 1 in 500 people.[13] That is less than 0.2%.

In 1914 only 15% of all heart disease was atherosclerotic in nature. Today that has risen to over 90%. The New England Journal of Medicine has stated, "The human genetic constitution has changed relatively little since the appearance of truly modern human beings…"[14] In other words, we are not much different genetically than the very first people who walked the earth. What is it, then, that has changed?

What has changed is our environment. Due to the industrial revolution we live in a more toxic world – toxic not only in the air we breathe, but most especially in the food we eat. While we have been led to believe that fat consumption is the biggest culprit in heart disease, the fact is that the only significant change in dietary patterns in Western countries over the last 100 years has been in refined sugar and flour intake, not fat consumption. This is significant because the environment in which our genes are bathed is what determines our genetic expression, meaning that approximately 75% of our genetic expression is more about what we *do* to our genes than the genes themselves.

Only 25% of our genetic expression, give or take, cannot be changed and is determined by the genes themselves. For example, you are not going to change your eye color from blue to brown with a change in your diet. However, since familial hypercholesterolemia affects only 0.2% of the population, chances are that if you have heart disease you are in the 99.8% of the people who can effect improvement by modifying the environment in which the genes are expressed. It is therefore possible for you to decrease your risk of heart disease by providing a better environment for your genes.

A better environment would most certainly begin with a better diet, as we will discuss in a later chapter.

CHAPTER 3

WHAT DRIVES CHOLESTEROL PRODUCTION?

Let's explore the role and benefit of cholesterol further, and the variables that can cause elevations in cholesterol.

So what drives cholesterol production? I can promise you it is not a statin drug deficiency! What *does* stimulate cholesterol production in the body is the blood sugar regulating hormone, insulin. The enzyme responsible for cholesterol production in the liver is the HMG CoA reductase enzyme, and insulin directly regulates it. Insulin is best known for its role in blood sugar regulation, but what is not well-known about insulin is its role its role in cholesterol production.

Insulin is produced in the pancreas and is secreted in response to dietary intake of sugars. Many whole foods have naturally-occurring sugars that cause a mild and temporary elevation in blood sugar, and that's healthy. But most of the processed foods available to us on the shelves of our local grocery stores have what is known as "simple" sugars – ones that get into the bloodstream very quickly and thus cause a spike in blood sugar and a subsequent spike in insulin. Since insulin that is chronically high can cause an unhealthy accumulation of fat in the blood and also on the body, it stands to reason that one of the primary culprits to consider in cardiovascular disease is sugar intake. Hence, sugar-laden snack foods, pastries, candy, soda, etc, are all the enemies of healthy blood sugar and a healthy cardiovascular system.

The prestigious medical journal, *The Lancet*, reported in 1996 that, "What has not been well recognized is that raised blood [sugar] concentrations much lower than those necessary to diagnose diabetes or even impaired glucose tolerance are also associated with an increased risk of cardiovascular disease (CVD). Thus, there seems to be a continuous relation between the risk of CVD and raised postprandial glucose concentrations that extends from barely elevated values right into the diabetic range..."

In other words, you don't have to have radically out-of-control blood sugar problems in order to have an increased risk of developing cardiovascular disease. All it takes to elevate one's risk to cardiovascular disease is blood sugar levels that are only slightly elevated.

Why is that? Probably the two primary reasons why this is the case is because elevated blood sugar stimulates insulin production. And when insulin is elevated for extended periods of time day after day, week after week, month after month, and year after year, it initiates a cascade of inflammatory signals coursing throughout the body. And we learned in the previous chapter that one of the major players in cardiovascular disease is inflammation. Likewise, insulin produces fat and can cause blood fats to rise. So the inflammation coupled with the elevated blood fats is the perfect storm that can lead to a heart attack. And this scenario cannot be relegated to only people in mid-life and beyond anymore.

The amount of sugar intake among children today is so immense that we can no longer legitimately call Type 2 Diabetes "Adult Onset Diabetes." A shocking number of children are being diagnosed with Type 2 diabetes due to the massive intake of sugar from the time of infancy. I have been shocked and outraged to observe parents allowing their toddlers to sip soda pop and Kool-Aid from bottles and sippy cups. It's no wonder that we are facing an epidemic of obesity, diabetes, and cardiovascular disease in people of younger and younger ages.

The Role of Stress

Another factor in cardiovascular disease is stress, which will actually cause cholesterol to rise in measurable amounts.

Do you know that your cholesterol production can go up over 100 points if you were to check your cholesterol prior to playing a vigorous game of basketball, and then coming back and testing your cholesterol again? Why is that? This happens in response to *a physically stressful environment*. The physical exercise is a form of stress, and the body must produce more stress hormones in order to compensate. And this rise in stress hormones requires readily-available cholesterol.

You may remember earlier I discussed that cholesterol is the precursor to the stress hormones. While we live in a society where we are no longer faced with stressors such as being chased by man-eating predators, our bodies nevertheless respond similarly to everyday stressors such as running late for work and being stuck in traffic. When stress hormones rise for *any* reason, the body must respond, and that response requires cholesterol.

The stress response allows the body to adapt to any physical, chemical, or emotional stress. Stress hormones are therefore necessary for survival itself. When the body needs more stress hormones in order to adapt to a stressful event, the body will increase the production of cholesterol, since cholesterol is the precursor to the stress hormones. If you have an emotional response that drives your stress levels up, for example, your cholesterol levels will go up as well. So if you get stressed out and you go and get your blood drawn, chances are that your cholesterol levels are going to be significantly elevated.

At the *Functional Medicine Research Center* in Gig Harbor, Washington, the research arm of the nutriceutical company, Metagenics, human trials are conducted year round. I recall being in attendance at a lecture where the Chief Science Officer of Metagenics, world-renowned natural medicine biochemist, Dr. Jeffrey Bland, was speaking. He explained that there was a lady participating in one of the human trials at the Center, and she had just had some blood work done and then promptly exited the building. Upon pulling out of her parking spot, she backed her car into a stationary object. Immediately one of the lab technicians ran out to ask if they could draw her blood again. And sure enough, her cholesterol was significantly elevated compared to what it was just a few minutes before. Why? Because she experienced something that initiated a stress response, and the production of stress hormones required that the body produce more cholesterol.

Another interesting observation is that in a laboratory setting machines are allowed to have a lab variance of 10% on either side in order for it to be considered an efficient machine to actually test samples?[15] For example, if your cholesterol comes in at 200, you could actually be anywhere from a 180 to 220. If the number is 220, most medical professionals would express great concern and say, "You're at 220! You need to be on a statin drug!" Unfortunately, that is what we are led to believe.

Sex Hormones and Cholesterol

Yet another critical physiological process that involves cholesterol is the production of sex hormones.

All steroid hormones are derived from cholesterol. Thus, any increase in the demand for sex hormones will cause an increase in the production of cholesterol. Think of middle-aged women. As their ovaries begin shutting down the production of sex hormones during perimenopause, the demand on the adrenal glands (stress glands) increases, which can cause a rise in cholesterol production. It's the law of supply and demand.

Toxicity

The *Journal of American College of Cardiology* says that lipoproteins (HDL, LDL, VLDL, etc.) provide **protection** to the body through **binding** and **detoxifying endotoxins** entering the circulation via the gut.[16] When toxins from the environment and/or our food enter the body, more cholesterol is produced in an attempt to bind to and eliminate those toxins.

In my practice I would routinely see patients' cholesterol drop by 100 points in four to eight weeks just by taking them through a nutrient-tailored detoxification program.

Cholesterol levels are indicators of compensating physiology and not necessarily the cause of cardiovascular disease, as we have been led to believe. Hopefully by reading up to this point you are beginning to understand that cholesterol is not evil.

But we're not done yet. Let's transition now and discuss cholesterol-lowering stating drugs in more detail.

CHAPTER 4

THE PROBLEM WITH STATIN DRUGS

Statin drugs have become standard procedure in the medical world, even if the cholesterol levels are healthy. I know some people who have heart catheters put in to open up clogged arteries who had cholesterol levels barely above 100, and yet they were put on statin drugs anyway!

Statin drugs artificially block HMG CoA reductase enzyme in the liver to lower cholesterol production. A drug like *Vytorin* blocks the enzyme production and also blocks the absorption of dietary cholesterol. The goal of this drug is to block the 75% production of cholesterol production in the liver and also block the 25% absorption of cholesterol in the intestinal system.

We know from the previous section that there are at least four things that increase cholesterol (insulin, stress, toxicity, and sex hormone production). Would taking Vytorin have any impact on the initiators of elevated cholesterol? The answer is no. This drug does not even touch the cause, because the causes are insulin, stress, toxicity, and sex hormone production. All the drug does is block the production in the liver and the absorption of cholesterol from the intestinal tract.

So would a patient who takes Vytorin be healthier or sicker? The answer is that he or she would be sicker. Why would it cause a person to be sicker? Because anytime a natural physiological process is inhibited or altered long term, most often than not there are going to be some consequences. If a pathway or mechanism in the human body is designed to function a certain way for the benefit of the entire body, and then we inhibit that function, it throws a monkey wrench in the gears, so to speak, and creates a cascade of events where the system begins to malfunction, often resulting in side effects.

The cholesterol-lowering statin drug, *Lipitor*, is currently the number one selling drug on the planet. Let's take a look at the background and what may have led to its widespread use.

In 1988, the *US National Cholesterol Education Program* stated regarding the treatment of high cholesterol, "Drug therapy is likely to continue for many years, or for a lifetime. Hence, **the decision to add drug therapy to the regimen should be made only after vigorous efforts at dietary treatment have not proven sufficient**"[17] (emphasis added).

Have you ever paid close attention to the commercials on TV marketing cholesterol drugs? There's a subliminal message that accompanies what they are directly telling you. They say, "When diet and exercise fail, try this drug..." In effect, what is emphasized, of course, is that you need the drug because diet and exercise are probably going to fail, and most likely already *have* failed in your case, even if you've never really given diet and exercise a try.

Consider, however, that a healthy diet and exercise are certain to fail if you don't engage in them for an extended period of time. I would be willing to bet that most people reading this book who are currently taking a statin drug were not seriously challenged and instructed by their medical physicians to undergo a vigorous lifestyle modification program for six months prior to prescribing your cholesterol-lowering drug. Most medical physicians are conditioned to whip out their prescription pads as first line therapy with any cholesterol reading over 200 and perhaps only briefly mention changing your diet as they hand you the prescription.

Yes, I am generalizing a bit, but not that much! There are some very caring medical doctors out there who will take the time to coach their patients on diet and exercise, but unfortunately those types of physicians are few and far between. The numbers don't lie. If more doctors were doing a better job of coaching their patients in the benefits of lifestyle modification, there wouldn't currently be 36 million Americans on statin drugs.

Let's go back again to the *US National Cholesterol Education Program*'s recommendations for prescribing cholesterol-lowering statin drugs. **"The decision to add drug therapy to the regimen should be made only *after* <u>vigorous</u> efforts at dietary treatment have not proven sufficient."** Alright then, what does 'vigorous' mean according to these guidelines? According to guidelines established by the NCEP, *six months of intensive dietary counseling* before starting drug therapy is the standard.[18] And yet the patients that are astute enough to know that their lifestyle is the issue often have to beg their doctors to let them have even three months before starting drug therapy.

Did you know that "all of the statin drug cholesterol trials were initiated <u>after</u> the publication of these guidelines, yet *none of the trials adhered to the 6-month dietary protocols* prior to putting subjects on statin drugs?"[19] Let me say that again. All the cholesterol trials involving statin drugs were performed after the publication of the US National Cholesterol Education Program's recommendations to administer statin drugs only after vigorous efforts at dietary treatment failed. And yet the researchers in the drug trials ignored these standards. Why do you think that is?

Why would none of the drug trials on statin drugs adhere to the six months of dietary protocols that were mandated? The answer is this: If all statin drug trial participants had been given dietary intervention before starting statins, *"...it would have much reduced the differences in deaths from coronary heart disease and all-cause mortality in the trials."* Essentially, dietary intervention would have skewed their stats, possibly causing statins to be deemed unnecessary.[20]

This is sad because of this summation: "Results of *several* studies have shown an *inverse relation*, or no relation, between total cholesterol concentration and risk of death in elderly people." One study "showed lowest survival rates for those with the lowest serum cholesterol concentrations."[21] So what does that tell us? In laymen's terms, it simply means that **lower levels of cholesterol usually result in people dying younger**! That's what the research says!

This same study showed that *"long-term persistence of low cholesterol concentration actually **increases risk of death**. The earlier that patients start to have lower cholesterol concentrations, the greater the risk of death."*[22] So the longer you keep that cholesterol count low, the more you're going to increase your risk of death!

There are eight-year-olds on statin drugs now because drug companies are pushing the administration of these drugs at that age for prevention! *Eight-year-olds!* I shudder to imagine what is going to happen to their bodies in the future when their sex hormone production is attempting to peak during puberty but the statins are interfering with that hormone production.

Even though the *American Academy of Pediatrics* recommends younger patients with elevated cholesterol readings focus on weight reduction and increased activity while receiving nutritional counseling, many young patients are administered the drugs instead.

Again I want to reiterate: **Lower cholesterol equals early death.** "The only significant overall effect of cholesterol lowering intervention that has been shown is to increase mortality."[23] Stop and think about that for a moment. Again, *the only thing that statins drugs have been shown to do besides lower your cholesterol is increase your risk of death!*

I'm going to run this one by you one more time just to make sure you got it, because I know it may be a hard pill to swallow given our current indoctrination. *The best clinical outcome we can get from cholesterol-lowering drugs is increased risk of death.* That is the *best* thing that we can expect, and yet 36 million Americans are on statin drugs![24] Are you kidding me?

Statins: Are they Friend or Foe?

The Lancet in 2007 stated that the number of Americans taking statins increases by 12 % every year. The same article stated that Harvard researchers analyzed the results of eight trials and discovered that the statins didn't reduce the number of deaths even among those with a known heart condition; and that statins may cause Parkinson's, and may be causing the heart conditions that they are supposed to be preventing.[25]

Remember the statistics about genetics and familial hypercholesterolemia? Recall that at best, genetics only applies to 1 in 500 people with cardiovascular disease. Yet in America we currently have 36 million people taking statin drugs to lower cholesterol. If we take 500 and divide it into 36 million, we have the figure 72,000. That means that of the 36 million people taking cholesterol lowering drugs, only 72,000 are taking it because of genetic issues. That translates to 35,928,000 people right now taking a cholesterol lowering drug for a *lifestyle* issue. Pretty sobering, isn't it?

Unfortunately for the majority of Americans, the true causes of high cholesterol and increased risk of heart disease are not being addressed. And this neglect by physicians in addressing lifestyle issues with their patients led to a staggering $27.8 billion in sales of statin drugs in 2006, half of that going to Pfizer, the maker of Lipitor.

Do you think the pharmaceutical industry wants to fix the true cause heart disease? Not with $27.8 billion at stake! Americans are easily convinced to take medication that promises a cure, but much to our disappointment, statins have done nothing to lower rates of heart disease in the last 20-plus years that they have been on the market.

Cardiovascular disease has been the number one killer for well over 20 years and it still continues to be number one today. However, I will point out that cancer is on the rise and will likely overtake cardiovascular disease soon as the number one killer. But as we will discuss later on, *statins have been linked to an increased risk of cancer as well.*

More Dirty Little Secrets about Statin Drugs

Statin drugs also lower levels of a very important nutrient called, Co-Enzyme Q10, otherwise known as simply, CoQ10. CoQ10 is a co-factor essential for the functioning of an enzyme system that creates 90% of the molecules needed for the production of the body's energy. It is produced by the same HMG CoA reductase enzyme that makes cholesterol.[26] By definition, *death is a lack of energy.* What have cholesterol lowering drugs been shown to do? *Cause death.* Part of the way they do this is by robbing the body of energy.

If 90% of the body's energy is supported by the production by CoQ10, and CoQ10 production is being decreased by statin drugs, we have a major problem. The organs that require more energy such as the heart have more of the energy-producing cells called, mitochondria. Therefore, these organs require *more* CoQ10 for their function.

In histology (the study of the microscopic anatomy of cells and tissues), the heart tissue is differentiated from skeletal muscle based on the increased number of mitochondria. The heart requires a lot of energy, as it is constantly beating with a large amount of force.

It's very interesting that congestive heart failure is linked to CoQ10 deficiency. Studies have shown that you can negate congestive heart failure with CoQ10 supplementation.

Again, the medical world relies heavily on drugs to treat cardiovascular disease, and yet too often these drugs are *causing* congestive heart failure! And what is congestive heart failure? It's *heart muscle cell death.* So in an attempt to lower cholesterol, the heart tissue is being damaged by the reduction of CoQ10.

The Journal of BioFactors stated in 2005, "We conclude that statin-related side effects, including statin cardiomyopathy (heart cell death) are *far more common than previously published and are reversible with a combination of statin discontinuation and supplemental coenzyme Q10.*" 27

It's clear that there are some major side effects from using statin drugs, but if patients stop taking the drugs early enough and begin supplementing with CoQ10, the side effects, including heart cell death, can be reversed!

If you are currently or previously a statin user, perhaps you noticed onset of muscle pain and fatigue while you were on the statin. Those side effects are common first-stage signs of CoQ10 depletion zapping the energy from the muscle cells. What is not as obvious, however, is how the statins also zap energy from the heart muscle as well, setting the stage for congestive heart failure.

So let's look at the actual statistics in the study mentioned above from the Journal of BioFactors.

Patients who stopped the statin drug and were given 240 mg of CoQ10 for 22 months had marked energy improvements and other benefits. Their fatigue went from a score of 84% to down to 16%. Myalgia (muscle aches and pains) went from 64% to 6%; shortness of breath went from 58% to 12%; loss of memory went from 8% to 4%. (Yes, loss of memory is also a common side effect of statin drugs because they deplete CoQ10 from the brain cells and can accelerate brain aging.) Peripheral neuropathy, which is the MS-type symptoms we talked about earlier, went from 10% down to 2%. [28]

As I mentioned in the beginning of the book, cholesterol is not bad in and of itself; it is *oxidized* cholesterol that is bad. Oxidation is the spark that ignites the fire that increases the level of inflammation in the body. This is important because as it has been stated, *"Atherosclerosis is clearly an inflammatory disease and does not result simply from the accumulation of lipids."* [29]

Quenching the inflammation that can lead to atherosclerosis can be as simple as ingesting essential fatty acids, like the EPA and DHA omega-3 fatty acids from fish oils, which can significantly decrease the inflammatory pathways in the body when in proper balance. However, Lipitor and other lipid-lowering drugs reduce the essential fatty acids that are clinically proven to prevent coronary heart disease and manage inflammation. [30, 31]

A 2010 article in *The New England Journal of Medicine* stated that people with Type 2 Diabetes have a considerably higher risk of cardiovascular disease, but "...*the cholesterol medications did not result in any notable differences in the rate of fatal cardiovascular events or the incidence of non-fatal first heart attacks.*" In this same article it was stated that "*previous research has shown dramatic reductions in heart problems in type 2 diabetics who combined medication with lifestyle changes such as exercise and eating five fruits and vegetables a day.*" [32]

Okay, so first we have them saying that the cholesterol drugs did not work when taken alone, but then they say that when they are combined with lifestyle modifications, such as diet and exercise, a dramatic difference was noticed. Translation: *The drugs don't work and lifestyle modification does!*

Believe me, there are reams of articles proving this point, but the political and corrupt nature of the whole cholesterol scandal restrains the doctor who wrote this article from being able to speak freely. He has to dance around the truth. In another place in the article he says, "*I'm not saying medicines aren't helpful, just that they need to be done in combination with lifestyle measures. It's reminding us how important our lifestyle is. You can't just overcome it with pills.*" [33]

I hope this point is abundantly clear. Drugs do not work to prevent or cure heart disease. They do not make you healthier, but they certainly do fatten the wallets of drug company executives. However, lifestyle modifications and specific supplementation *has* been shown in the research to do what the drugs cannot.31

CHAPTER 5

THE WOLVES GUARDING THE CHICKEN COUPE

Given all this shocking information about the danger of statins, it leads one to the obvious question as to why they are still available on the market. It does not take a detective to uncover the truth.

When the *National Cholesterol Education Program* (NCEP) lowered the "optimal" cholesterol levels in 2004, **8 out of 9 people on the panel had financial ties to the pharmaceutical industry.**[34] What could be gained by lowering the standard accepted level of desirable cholesterol? It is sad but probably true that the panel made this recommendation so that more people would be on cholesterol lowering drugs purely for their own financial interests.

In addition, statin drug researchers are similarly well-compensated. This quote offers some insight:

"It's almost impossible to find someone who believes strongly in statins that did not get a lot of money from the industry." [35]

Cholesterol lowering drugs may be taken for over 30 years by most users, and the FDA has approved these drugs based on clinical trials that have lasted only a fraction of that time. And even in the short duration of time of these trials, the drugs have shown to only increase the risk of death to users, albeit with lower levels of cholesterol. Millions of people with no symptoms of heart disease are now taking these medications, not knowing that the ultimate effects are unknown.[36]

Here is an example of the misleading advertising for statin drugs. You may recall the Lipitor television ad where Dr. Jarvik (maker of the artificial heart) says, "When diet and exercise aren't enough, adding Lipitor significantly lowers cholesterol." Then they show a statistic on the screen that Lipitor reduces the risk of heart attack by 36%. This is a very misleading statistic.

Here is a simplified version of the study: A group of 100 people took statin drugs for two and half years and two of them had a heart attack. In the placebo control group, a group of 100 people took a dummy pill (placebo) during that two and half years and three of them had a heart attack. So the statisticians say 2 out of 100 versus 3 out of 100 is a reduction by 1/3. Since 1/3 is 33%, that was used as a relative statistic advertised. There are decimal points and other details that cause them to push the number up a little to 36%, but that is beside the point. The point is that *it's a relative statistic; it's not a real statistic.*

Let's break this down in attempt to understand a relative statistic versus a real statistic.

There were 2 people who had a heart attack out of 100 in the group receiving the statin drug, and 3 people out of 100 who had a heart attack in the other group receiving a placebo. Is that really a significant difference? No. In fact, those results are virtually meaningless.

In fact, *the Journal of the Royal Society of Medicine* (2004) stated that the absolute *best* benefit that can be expected from a statin drug is a difference of only 1% to 3%. They also state, "These are not impressive results."[37]

In reality, if one analyzes the research it becomes clear that the chances of reducing heart disease with cholesterol-lowering drugs are as likely as the chance of someone winning the lottery. The actual benefit derived from these drugs is almost non-existent.

There is a term in drug research called the NNT, which refers to the "Number Needed to Treat." For example, if you give an antibiotic to 11 people with an H. pylori infection in their stomach (an infection that can cause ulcers), 10 of them will get better. The NNT of that is 1.1. So it is almost a 100% probability that the antibiotic is going to get rid of the infection. On the other hand, "Anything over an NNT of 50 is worse than a lottery ticket; there may be no winners."[38]

The studies with the best NNT for a statin drug that I could find were 100. That is correct! An NNT of 100 is more than *double* the likelihood of winning the lottery! The worst I found was 250. That's more than *five times* the likelihood of winning the lottery!

Drug industry critic Dr. Jerome R. Hoffman, professor of clinical medicine at the University of California at Los Angeles, summarizes how an NNT of 250 translates in the real world.

"What if you put 250 people in a room and told them they would each pay $1,000 a year for a drug they would have to take every day, that many would get diarrhea and muscle pain, and that 249 would have no benefit? And that they could do just as well by exercising? How many would take that?"

Looking at it from that point of view is eye opening. It is unlikely that there would be a single person who would agree to take the drug in that scenario. Unfortunately, due to mainstream marketing, 36 million Americans are doing this every day without the benefit of knowing what you have now learned.

The Purpose and Actual Benefit of New Drugs

When the patents on various drugs expire, companies scramble to make alternatives that will replace the revenue that will be lost. But independent reviewers found that about **85 percent of new drugs offer few, if any, new benefits, but they carry the risk of causing serious harm to users.**

In this scramble to find new drugs to market, many drug manufacturers begin experimenting with mixing pre-existing drugs together. Remember the drug, Vytorin, the drug that blocks the absorption of cholesterol in the gastrointestinal system and blocks cholesterol production in the liver? Vytorin is a combination of two pre-existing drugs: Zetia, and an earlier statin called, Zocor. It produced over $2 billion in sales in the first year. "It reduced LDL's more than a statin alone…, but *the patients' arteries thickened more* when taking the combination than with the statin alone and the drug *didn't bring any added benefits*." [39]

In other words, the drug did not offer any new benefits over other drugs, but what it *did* do was cause hypertrophy (thickening) of the smooth muscles of the arteries. This may actually accelerate the heart disease it was supposed to be preventing by narrowing the arteries, causing high blood pressure!

What happens in fluid dynamics when you make a hole narrower? If you were trying to physically blow water through a narrow hose, that task is considerably more difficult the narrower the hose. The pressure is higher in a more narrow hose, and therefore it takes more force to blow the water through it. The same thing happens if the length of the hose is increased significantly. The longer the hose, the more pressure or force is needed to blow the water through it.

Researchers proved that using Vytorin actually narrowed the blood vessels, which can increase the pressure in the blood vessels, which is called *High Blood Pressure*, a major risk factor for cardiovascular disease.

Without this knowledge, a patient using Vytorin may be put on a blood pressure lowering drug to counter the increased blood pressure, and the cycle of symptom-chasing begins – a cycle that ends with the average person over the age of 65 being on 5 or more prescription drugs, the majority of which are not actually addressing the true cause of their symptoms.

Here's an example of the importance of addressing the underlying cause.

Recall that we discussed the role of insulin in a previous chapter. Insulin rises when there is sugar in the bloodstream. Insulin helps the body store that excess sugar in the fat cells of the body. Remember that we discussed how insulin increases cholesterol production as it stimulates the HMG CoA reductase enzyme. Insulin at above-normal levels will produce fat inside the body in the form of blood fats leading to arterial plaque when oxidized, and it will also produce fat on the body in the form of weight gain. This is important to consider since **every pound of extra fat brings with it an average of 200 miles of extra blood vessels**[40] **that will further increase the blood pressure**. In this case, the underlying cause is the junk food consumption that increases insulin release. The insulin increases the fat production causing an increase in the amount of blood vessels, thereby increasing the resistance and leading to the high blood pressure.

As you can see, giving a blood pressure pill does nothing to fix the cause of the problem, and by not treating the true causes of a condition the patient only becomes more unhealthy over time.

Let's look at Vytorin again, but from another angle. "A five year trial did not show a reduced cardiovascular risk, but **a large percentage of patients treated with Vytorin were diagnosed and died from all types of cancers combined** when compared to the treatment of a placebo."41

I encourage you read that last paragraph once more. The *patients died from cancers!* The Lower the LDL's, the greater the risk of cancer.

*"LDL level of 107 was associated with a 33% increased risk of cancer and death, an LDL level of 87 was associated with a 50% increased risk of cancer and death. As the LDL goes lower the risk keeps getting worse."*42

Additionally, *The Journal of the American Medical Association* stated way back in 1996 that **"most cholesterol lowering drugs cause or promote cancer."**43 Researchers and drug manufacturers have known that statin drugs cause cancer since 1996! Yet we still have 36 million people taking statin drugs today, and drug manufacturers are still heavily promoting them because they are huge money-makers.

Indeed, the data showing cancer-causing nature of these medicines has been published for many years and submitted to the FDA. How is it, then, that cholesterol-lowering agents were approved by the FDA for long-term use in spite of their carcinogenicity in animal studies? In answering this question let's take a peek behind-the-scenes of the Endocrinologic and Metabolic Drugs Advisory Committee meetings at which the cholesterol drugs lovastatin (Mevacor) and gemfibrozil (Lopid) were discussed. (This information is made available by under the Freedom of Information Act.) The only reported discussion of the revelations about the cancer-causing nature of these drugs in the studies at the FDA Advisory Committee meeting on lovastatin (February 19th and 20th, 1987) was by a representative of Merck, Sharpe and Dohme (makers of the Mevacor brand of lovastatin), and it is no surprise that they downplayed the importance of the studies.

At the end of this meeting, a vote was taken by the Advisory Committee concerning the safety of Lopid (gemfibrozil) and the minutes state that only **"three of the nine members believed that the potential benefit of using gemfibrozil for prevention of coronary heart disease outweighed the potential risk associated with such use."** Unfortunately, such votes are only "advisory" to the FDA, who then decides whether to approve or deny a drug based upon "other information." However, it is this "other information" that is produced by the very company that manufactures the drug in question. In the case against Lopid, the FDA went ahead and approved it for the prevention of coronary heart disease despite the vote of the Advisory Committee. This is madness! But this is not an uncommon practice, unfortunately.

As my friend, Andrew Robbins, put it in his excellent book entitled, *Licensed to Kill: The Growing Epidemics of Iatrogenic Disease and Bureaucratic Madness*, there is surely going to be an especially hot place in hell for these money mongers who sacrifice human lives on the altars of their gods of money and power.

In 2005, *The American Cancer Society* published an article stating that 175 people are diagnosed with cancer in America *every hour*, and 65 will die from it. They added that two thirds of those people could be cancer free if they ate right, exercised, did proper screenings, and quit smoking. But based upon the research above, we could also add that many of these might be cancer free if they stopped taking their statins!

Statins Sell Erectile Dysfunction Drugs as Well

Statin drugs are not only huge sellers themselves, but it turns out that they also provide a platform for the sale of other popular drugs not directly related to treating high cholesterol.

For example, Pfizer's number one selling drug is Lipitor. Can you guess what Pfizer's number two selling drug is? *It's Viagra!*

Released in March 1998, Viagra had the most successful first year of any pharmaceutical ever launched, reaching $1 billion in sales on the first anniversary of its launch. By the end of 1998, sales of the pills, costing $7 each, amounted to $656 million in the USA alone. Total sales, including international, was

$788 million. In 2005, it was estimated that 23 million men had received Viagra prescriptions, with global annual sales amounting to 1.6 billion dollars.

Is there a cause and effect at work here? Let's find out if there's a connection between Lipitor and Viagra.

Earlier we learned about the important roles that cholesterol plays in the production of our sex hormones such as testosterone, progesterone and estrogen. So now let's test your memory by asking a question: *How does cholesterol get to the testicles (gonads) in order to be converted into testosterone?*

The answer is that cholesterol gets bound to the LDL cholesterol and transported to the testicles (gonads) where the testosterone production takes place.

What happens to the LDL levels when you take a statin drug? LDL levels go down, right? Thus, how will cholesterol reach the testicles to be converted into the sex hormone testosterone if you lower your LDL levels? The answer is that cholesterol's transport to the testicles will be inhibited, and that may lead to a decrease in the production of testosterone.

Testosterone has many effects on the body, but it is particularly involved in libido and penile erection frequency. So what can happen if you lower your LDL levels leading to a possible decrease in testosterone production? Males are going to be more predisposed to erectile dysfunction, low or no sex drive, lethargy, etc.

To show this is not simply my theory, in 2010 *The Journal of Sexual Medicine* published a study that evaluated nearly 3,500 men who had erectile dysfunction (ED). They stated, "Statin therapy prescribed to lower cholesterol also appears to lower testosterone. Current statin therapy is associated with a two-fold increased prevalence of hypogonadism, a condition in which men don't produce enough testosterone."[44]

What a lucrative business model! Create a statin drug that rakes in billions of dollars annually, which creates the need for a second drug to counter the sexual side effects, which is also a blockbuster of similar magnitude. It must be good to be a pharmaceutical executive….that is, as long as you can sleep at night!

If statin drugs can make that much of an impact on the sex hormones of adults, imagine what they are doing to the 8-year-old children who are on them. Again, I can only imagine the developmental issues that may arise as children on these drugs reach the age of puberty. I have to imagine, because there are no studies on the long-term effects of children taking statin drugs. However, looking at the basics of human physiology, there is enough information to tell me that the outcome will not be positive.

Make no mistake about it, statin drugs negatively affect everyone who takes them—men, women, and especially the children. Although "simple blood tests" may not yet be showing that one's liver enzymes are elevated, indicating liver cell death, those tests are by no means proof that the statins are not harming the body in other ways.

Dear reader, please consider the proof and make your own decision. These drugs have been proven to cause cancer and congestive heart failure and sexual side effects. They have been clearly demonstrated to lead to early death. So are they still on the market for health reasons, or for financial ones? I'll let you be the judge.

In closing this chapter, I also want to encourage you take inventory of your own lifestyle habits. Granting some exceptions, the cause of disease is due, by in large, to our own lifestyle choices. The statistics show that we must look in the mirror and stop blaming genetics or waiting for the next miracle drug. We *must* change the way we eat, the way we think, and the levels exercise we get in order to enjoy life to the fullest. More on this in a later chapter.

CHAPTER 6

DEFICIENCIES CAUSED BY CHOLESTEROL LOWERING DRUGS

Up to this point we have only talked about statin drugs, so let's transition now and discuss another category of drugs that doctors are prescribing to help lower cholesterol. Many people are being prescribed a category of drug called *fibrates*. Fibrates reduce the production of triglycerides and can increase HDL cholesterol. But this can be accomplished simply by exercising and eating a healthy diet low in processed sugar. There are no side effects to a healthy lifestyle, just side benefits! However, the side effects of fibrates are mild upset stomach, myopathy (muscle cell death), and increased cholesterol content of the bile, which increases the risk for gall stones.

When fibrates are taken with the statins there is an increased risk of muscle cramping, myopathy, and of rhabdomyolysis. Rhabdomyolysis is the breakdown of skeletal muscle tissue which can lead to kidney problems. Additionally, fibrates deplete the body of Vitamin B12, Vitamin E and CoQ10. [45, 46, 47]

Another category of cholesterol lowering drugs are the bile acid sequesterants. Bile sequesterants prevent reabsorption of cholesterol from bile in the gut.

Bile is a digestive juice secreted by the liver and stored in the gallbladder and aids in the digestion of fats. Since bile acids are synthesized from cholesterol, the disruption of cholesterol absorption would decrease cholesterol levels. But not only do they decrease the absorption of cholesterol, they decrease a number of vitamins from being absorbed, especially the fat soluble vitamins like A, D, E, and K.

In addition to the fat soluble vitamins, sequesterants also deplete the body of B12, calcium, magnesium, phosphorus, zinc, iron, folic acid, beta carotene, and the dietary fat that we consume.[48]

As we learned earlier, when fat is added to the diet in large amounts, the cholesterol production by the liver is halted. And when we remove dietary fat the body's cholesterol (blood fat) production goes up. So by depleting the body of dietary fat through the use of sequesterants, all that does is cause cholesterol production to go into overdrive. It has also been documented that by inhibiting the absorption of fat in the gut, more thyroid hormone (Thyroxin or T4) will be lost in defecation, which lowers the thyroid hormone levels (Thyroxin or T4 levels). So now these patients are put on yet another drug for low thyroid.[49]

Can you see why so many people are on prescription drugs and why they never get healthier?

So let's analyze in more detail some of the specific nutrients that are depleted by these drugs and why they are so important for cardiovascular and overall health. As we take a look at the significance of these deficiencies, you will see yet another reason why the drugs that are being used for the prevention of cardiovascular disease are actually increasing the risk of it.

Vitamin B12

Vitamin B12 has many biological functions, including conversion of RNA to DNA (replicating the genetic code), synthesis of myelin (maintenance of nervous system), synthesis of methionine and metabolism of folic acid, maturation of red blood cells, and metabolism of protein, fat, and carbohydrates.[50]

We have already learned that insulin and carbohydrate regulation is key to maintaining a clean bill of health with regards to heart disease. However, a deficiency in B12 also decreases the presence of folic acid, which in turn increases the production of homocysteine, a potent free radical (causes increased oxidation). This leads to increased inflammation, an underlying cause of heart disease.

Folic Acid

As discussed above, when folic acid metabolism becomes faulty, it can lead to an increased production of homocysteine. Research has confirmed that high homocysteine production is a major contributor to heart disease because it is a potent free radical and contributes to inflammation.

Folic Acid has a number of other biological functions as well, including synthesis of RNA and DNA (essential for proper cell division and transmission of genetic code to new cells), prevention of cancer, prevention of birth defects, and healthy maturation of red and white blood cells.[51]

Vitamin E

Vitamin E is an antioxidant. It prevents free radical damage and protects LDL-cholesterol from oxidation, thus protecting blood vessels from atherosclerotic lesions (plaquing in the arteries). It also decreases platelet stickiness.[52]

That means that instead of pushing blood resembling the consistency of syrup through your veins, vitamin E makes the consistency of the blood thinner and easier for the heart to pump throughout the body. This decreases the pressure in the blood vessels (lowers your blood pressure). Thus, when you are deficient in Vitamin E, all risk factors of heart disease increase.

CoQ10

The symptoms of a CoQ10 deficiency are as follows: high blood pressure, angina, mitral prolapse, stroke, heart arrhythmias, cardiac myopathy, congestive heart failure, poor insulin production, lack of energy, gingivitis, and weakening of the immune system. You can see the majority of signs and symptoms are related to cardiovascular function.

As mentioned earlier, CoQ10 is involved in energy production, but it also acts like an antioxidant, preventing LDL-cholesterol oxidation, thus reducing cardiovascular disease and atherosclerosis risk.[53]

Beta Carotene

Beta carotene quenches free radicals, thus decreasing oxidative damage, which is the initiator of inflammation—the major cause of heart disease.[54]

Calcium

When you have deficiency in calcium it can cause heart palpitations and high blood pressure, further increasing the risk of heart disease. Calcium also has a number of biological functions that are too numerous to list here in their entirety, but here are a few: Calcium supports development and maintenance of healthy bones and teeth, initiation of muscle contractions, regulation of fluid passage across cell membranes, activation of enzyme systems responsible for muscle contraction, fat digestion, and protein metabolism.[55]

Iron

It is not uncommon for iron levels to get low enough to cause iron deficiency anemia (low red blood cell count), which has been linked to cardiac myopathy, angina, and heart attacks. Iron has many functions. A main function is oxygen transport by hemoglobin, which means that it picks up oxygen in lungs and transports it to tissues. Iron is necessary for optimal immune response, and supports the synthesis of the amino acid, L-carnitine, which helps the metabolism of fat. Iron also has a role in energy production, liver detoxification enzymes (enzymes that remove toxins from the body), and is a component of enzymes that initiate synthesis of serotonin and dopamine, the feel good chemicals in the brain. Additionally, iron is involved in the synthesis of collagen and elastin, which makes your skin, ligaments, and blood vessels stretchy and pliable.[56]

Magnesium

Magnesium deficiency can cause high blood pressure and cardiac muscle spasms resulting in a heart attack.[57] Magnesium is a smooth muscle relaxant. A deficiency in magnesium can cause constriction of the smooth muscles that line the blood vessels, which increases the resistance. This will, of course, lead to high blood pressure.

A biological function of magnesium is decreased platelet aggregation.[58] In other words, magnesium decreases the thickening of the blood. Again, if the blood is thick like syrup, the blood flow resistance increases, thus leading to increased blood pressure. If the blood is thinner like water, there is less resistance to blood flow, thus lowering blood pressure and a decrease in the risk of heart disease.

Potassium

Potassium, also depleted by bile sequesterants, has many functions, including but not limited to controlling distribution and balance of water in the body, conduction of nerve impulses, contraction of muscles, and maintenance of normal cardiac rhythm and of pH (acid/base) balance in the body.[59] Acid base balance in the body is very important because every enzyme and every hormone and their respective receptor sites are all pH dependant.

Vitamin D

Last but not least on the list of nutrients depleted by heart drugs is Vitamin D. Vitamin D deficiencies have been linked to *heart disease, hypertension*, autoimmune diseases, certain cancers, depression, chronic fatigue and chronic pain.[60]

This is an incomplete list, but I believe you get the point. Nutrients are absolutely vital for life and optimum health, and drugs deplete many of these life-giving nutrients.

At this point, perhaps you should pause for a

moment to think about all you have been reading about up to this point. I know that probably most readers may have never been exposed to much of the information you have learned here, and you may be asking why you have not been told this before. The sad but true answer is that it's all about the money.

Let's be honest. Would you have taken any of the cholesterol lowering drugs if you knew this information before? Not likely. If the masses had the information readily available you have been exposed to here, many drug companies would be out of business, and many doctors would be as well. They are not going to let that happen if they can help it.

So now you might be asking yourself why the pharmaceutical industry would lie to you. They are here to help us, right? Wrong. There is a place for drugs in certain situations, and in the case of antibiotics, for example, certain categories of drugs have been life savers. But we also have to realize that the pharmaceutical industry is a *business* that exists to make money, and we now know they would definitely lie to us in their marketing. Pfizer, the maker of the number one selling drug in the world, Lipitor, got slapped with a $2.3 billion lawsuit for *illegal marketing methods*! That means they lied! But do you think that really hurt them, considering that they made half of the $27.8 billion from total statin sales in 2006 alone? I don't think so. With sales numbers like that, a $2.3 billion lawsuit is barely a slap on the wrist.

It's time for us, the American consumers, to wake up! We are the victims of conspiracy, fraud, and deception by the very ones who are supposed to be protecting our health interests.

CHAPTER 7

LIFESTYLE AND DISEASE

Up to this point I realize that it may appear that I am anti-medicine. I want to make it clear, however, that I'm not anti-medicine, I am simply pro-truth and pro-health. I believe there is a time and place for traditional medicine, and that is in emergency acute care situations. America has the finest emergency medical system in the world, and we excel in crisis care. However, that cannot be said of our chronic care methods. Most of the time a more effective treatment for chronic disease is a change of lifestyle, not another prescription.

To illustrate my point, imagine your house catches on fire. Who would you call to deal with that crisis? You would call the fire department, right? So the fire department arrives with their axes and hoses, and they break down the door and hose everything down. And in doing so they save you and your house by extinguishing the blaze.

The next day you realize that while your house has been saved, you still have an enormous mess on your hands. You have to replace the windows, tear out the charred drywall and replace it with new, repaint, re-carpet, and so forth. And who do you call for that job? Do you call the fire department back with their axes and hoses to rebuild your house? No. Why not? Because they don't have the right tools to rebuild the house! You need someone with the appropriate set of skills and tools.

The western model of medicine is like the fire department. They excel in addressing sudden and life-threatening crisis situations. But addressing the arduous task of chronic disease and ongoing wellness is not their forte, and the current statistics on chronic disease bear that out. A Surgeon General report confirmed that 7 out of 10 leading causes of death in America are diet and lifestyle related, and western medicine has not made any difference at all in improving those numbers. Rates of chronic disease like cancer and diabetes continue to skyrocket unabated by the methods of western medicine.

As it pertains to the primary subject of this book, cardiovascular disease is one of those lifestyle diseases referred to in the Surgeon General report. If you want to decrease your risk of heart disease, you must change your lifestyle, because there is no magic pill to circumvent an unhealthy lifestyle. You also need a healthcare practitioner with a different set of skills and tools who can help you change the way you think about your body and how it functions. You need a handyman, so to speak – a doctor who focuses on health and wellness and can help you rebuild your house (body); not a doctor who just puts out fires by giving a pill for an ill and chasing symptoms.

One major problem with our current health care system is that we have a bunch of firefighters coming in to rebuild houses and do repair work with the same set of tools they use to put out fires, and it's not getting the job done. We continue to limp along with this failed mentality, hoping at some point that perhaps the outcome will be different. Albert Einstein said it best by declaring, "One definition of insanity is doing the same thing over and over and expecting different results." We continuing on this path at the expense of millions of people who have put their faith into what they think is health care, when in reality it is not *health* care at all; it is disease care.

I quoted an article earlier that two-thirds of cancer victims could be cancer free if they ate right, exercised, quit smoking, and received proper screenings. Yet we spend millions of dollars on chemotherapy agents to save 1 person out of 100, when we could be saving closer to 70 out of 100 if we would teach and help people to be healthy. The saying "knowledge is power" is true, but an even truer statement is that knowledge *acted upon* is power.

How can we fix the problem of heart disease? By finding the cause. What drives cholesterol production? Insulin, toxicity, sex hormone production and stress drive cholesterol production. More importantly, what makes cholesterol bad? Oxidation that leads to inflammation is what makes cholesterol bad no matter what the cholesterol levels are, and that's information you can bet your health on.

Functional Medicine to the Rescue!

Functional Medicine is a term used to define a different method of treatment and diagnoses which focuses on discovering and treating the underlying cause of disease. Functional Medicine in its simplest form is discovering the physiological links of cause and their subsequent effects (signs and symptoms).

Typical western medicine tends to treat symptoms, which are simply manifestations of a biochemical imbalance or a physiological malfunction. Functional Medicine, on the other hand, analyzes the entire body physiology to focus on the causes of illness and attempt to restore optimal health. Instead of isolating one ailment, Functional Medicine works from the premise that at our basic core structure we are nothing more than trillions of cells floating around, much like fish in a fish tank.

If the water in a fish tank became polluted and the fish got sick, you would not take the fish out of the water and try to address the illness with medication. You would simply replace the water in the fish tank. In the same way, the health of our cells is determined by the quality of the environment in which they are living.

Functional Medicine practitioners utilize time-honored methods as well as breakthrough technologies to assess and treat illness. In some cases, various body biomarkers are analyzed with new emerging laboratory technology that can help to determine the health of the cellular environment. Once these assessments are made, functional medicine practitioners will typically recommend a regime of specific dietary modifications, various types of exercise, and the use of nutriceuticals (natural alternatives to drugs) in order to restore proper health and balance of the cellular environment. When the cells are in a healthy environment, thus empowered to function properly, health is restored and illness is eradicated. But this takes time.

To experience lasting change in one's health, one must be committed to three things primarily: frequency, duration, and intensity. The more frequently you do something, the longer you do it, and the more intensely you do it will determine how fast and how profoundly changes will take place.

The Story of the Perfect Ten

To illustrate my point that it takes time to regain health, here is the story of what I call the "Perfect Ten."

Let's imagine that the health of our cells can be measured on a scale of one to ten, one being the least healthy and ten being the most healthy. Using this scale as a measuring stick, let's consider what happens during cell replication.

When our bodies first begin developing at conception, the very first thing that happens is one cell divides. That very first cell, under normal circumstances, is a perfect ten on the scale of health. When that first cell divides, what exactly does it do? It creates an identical replica and that replica becomes another healthy "ten." And those cells carry on that same process of replication and continue multiplying, and eventually that replication mysteriously begins to specialize into cells for the eyes, cells for the hair, cells for the skin, and so on until a perfect little baby is ready to be born.

Once that precious little baby is born, the modern American method of caring for it is to begin feeding it baby formula, because we apparently think a synthetic man-made powdered drink mix is better than God's perfect design of mother's milk. (Of course, I understand there are times when parents adopt and have no other choice but to use formula, or perhaps a mother cannot produce her own milk and must resort to formula, but that is beside the point.)

My point is that when we put man-made compounds into the body, the body cannot function at its optimum, and health is compromised ever-so-slightly at first. So when cell replication occurs again, our cells are no longer a "Perfect Ten," but are functioning as perhaps a seven or an eight.

Pretty soon we decide the child is now at the age where he/she can have Kool-Aid, as if taking two cups of sugar and food coloring, putting it in two quarts of water and giving it to our kids is a good idea. So now let's say that at this point the cells are functioning at a lower level, perhaps down to a five or six. Perhaps our children are getting overweight or are sluggish or have attention deficit.

As they get a little older, we loosen up our standards even more and allow them to have diet soda pop. And what is the sweetener in diet sodas? *Aspartame.* Aspartame once ingested breaks down to formaldehyde, which can wreak all kinds of biochemical havoc.

(When I did my fellowships in anatomy I had to work in an environment where there was formaldehyde. I had to wear waders. I had to wear a respirator. I had to wear gloves. I had to be completely covered head to toe. Why? Because formaldehyde causes cancer. Although most people think they are doing a good thing to the body by drinking diet sodas sweetened with aspartame because they are not getting the sugar, they don't realize they are ingesting a chemical that breaks down in the body into cancer-causing compounds.)

To get on with our "Perfect Ten" analogy, once living on diet soda and junk food for several more years, now the cellular health may have diminished down to a four. When those cells replicate, what happens? The cells divide and make another set of cells that are also functioning at a four. At this point we have been conditioned to break out the medications that make us feel like a ten for a while because *we're inhibiting altered functions, thus decreasing the symptoms.* We may feel like a ten initially, but are we healthier? No, because in reality our cells are still functioning at a four, if not worse. What we really need to be doing is fixing our baseline by changing the environment our cells are bathed in on a daily basis.

That is accomplished by changing the quality of the fuel. Whole, unprocessed food is the order of the day. As the cells are saturated in this new healthy environment, they begin to function better, and when they replicate they may regenerate back up to a level six. As we continue giving the cells more nutrients, their function continues going up to a seven. When those cells replicate, they produces more level "seven" cells.

Over time as we continue this lifestyle and experience the benefits in how we feel, we are inspired to continue the process of regaining our health and vitality and eventually getting all the way back up to a "Perfect Ten," as we were meant to be.

CHAPTER 8

MAPPING OUT A COURSE TO BETTER HEALTH

If you have stuck with me this long and read this far, perhaps you are now ready to take inventory of your own health and lifestyle, and are inspired to make some changes. If so, that's wonderful, and I applaud that. But don't make those decisions hastily and without counting the cost, because repairing the damage and getting back on track to vibrant health through a new healthy lifestyle is not a quick fix. Change is not going to happen overnight. Our bodies do not become unhealthy and degenerate into chronic disease in a day, and therefore it is not reasonable to expect the road back to health to always be an easy one.

Certain types of medications may offer some symptomatic relief, but the truth is, medications don't make a person healthier because drugs do nothing to make the cells function more efficiently. Renewing cells that are functioning on a level four back up to a level nine or ten so that you don't need drugs to mask symptoms takes a consistent effort for an extended period of time, and that timeline is different for each person because of different life circumstances and genetic uniqueness.

As with any journey that a person sets out on, one important tool for reaching your destination is a roadmap. In this case, your roadmap toward optimal health would be discovering what kind of condition your cells are in as you get started so that you know where you want to go and what it will take to get there.

In helping you to map out your course, then, I would like to provide a quick summary of what I believe are some of the most important biomarkers to assess your internal environment regarding optimal cardiovascular health. There are many doctors that can help you identify these biomarkers, and the appendix section of this book provides a list of sources to help you locate practitioners in your area who provide this kind of assessment and care.

Oxidative Stress

As mentioned earlier, free radicals are a byproduct of oxidation, or the process of metabolizing oxygen. An example of oxidation is the browning of fruit or the rusting of metal. Oxidation is a normal body process and necessary for certain functions, but high amounts of free radical production accelerates cellular aging and degeneration, and thus makes the body more susceptible to illness. The body counters oxidation by producing and ingesting antioxidants, the molecules that act to neutralize free radicals.

Antioxidants protect every cell of the body against damage caused by free radicals, and a lack of antioxidant reserves and/or an over-abundance of free radicals can lead to all manner of dysfunction, including (but not limited to) fatigue, weakness, headaches, compromised lymphatic function, cognitive impairment, inflammatory conditions, accelerated aging, infections, respiratory disorders, arthritis, autoimmune diseases, increased risk of cancer, and increased risk of cardiovascular disease.

Inflammation:
The Consequence of Excess Oxidation

Chronic inflammation is becoming known as the "engine that drives many of today's most feared chronic diseases."[61] Inflammation is a natural process in response to injury or degenerative changes; and when it becomes excessive and chronic, it can have devastating consequences.

Regarding cardiovascular health, the amount of cholesterol a person has is not so much the problem; the amount of inflammation coursing through a person's body is the bigger problem because that inflammation can cause cholesterol plaque to build up in the arteries and trigger a heart attack or stroke. However, due to imbalances in the body, genetics, and lifestyle issues, inflammation can become chronic in a low-grade state, meaning that it usually goes undetected and misdiagnosed.

Liver Stress and Toxicity: A Major Cause of Oxidation

Toxicity is a double-edge sword when it comes to cardiovascular health. First, it's detrimental because it significantly increases oxidation and free radical activity throughout the body, and that is bad for the cardiovascular system for reasons that we have already discussed. Second, in our day and age, toxins in the environment and in our food supply are so abundant that the body cannot keep up with detoxifying them and eliminating them properly.

When toxins accumulate throughout the body, fat cells will likewise accumulate because the body will use fat cells to store the toxins in an attempt to prevent them from damaging the vital tissues and organs. As fat accumulates *on the body*, it also accumulates *in the body* in the form of blood fats (cholesterol). These two detrimental effects can lead to the formation of plaque in the arteries, which will cause heart disease.

Additionally, when the toxins accumulate within the body organs and tissues, it leads to many chronic illnesses such as fatigue, neurological disorders, various cancers, allergies, problems with blood sugar metabolism, chronic headaches, inflammatory and autoimmune disorders, atopic disorders (disorders of the skin), sinus problems, digestive disorders, anxiety and depression, immune dysfunction, and hormone metabolism problems, to name just a few. Thus, nutrient and diet-tailored detoxification can be important in eliminating the cause for various illnesses.

Carbohydrate Metabolism

When dietary carbohydrates are ingested, they are converted into a sugar (glucose) that is dumped into the bloodstream. The pancreas then responds by secreting the hormone insulin, which stimulates the liver and muscles to absorb the glucose and convert it to glycogen (glycogen is stored sugar). Because of excessive amounts of simple carbohydrates in the typical American diet (the average American consumes a whopping 158 pounds of sugar per year),[62] the insulin receptors on the cells can become desensitized to insulin. This causes inefficient glucose utilization, which leads to fatigue, carbohydrate cravings, weight gain, neuropathies, cardiovascular and neurological issues, and ultimately diabetes as the pancreas secretes larger and larger amounts of insulin in an attempt to compensate for the un-responsive cells.

As stated earlier, insulin directly stimulates the HMG CoA reductase enzyme, thus increasing the amount of cholesterol present; and the increased amount of simple sugars increases the amount of oxidation and free radical damage, thereby increasing their risk for cardiovascular disease.63 Identifying problems in carbohydrate metabolism is one of the most important factors in overall health. The importance of this aspect of our health is underlined by the fact that 36 cents of every healthcare dollar is spent on issues related to blood sugar metabolism, and these issues lead to the number one killer of Americans – cardiovascular disease.

Adrenal Stress

Adrenal stress is an important consideration in recognizing the etiology of various types of disorders since many have been known to be associated with adrenal dysfunction. Adrenal stress can have a negative impact on blood pressure, blood sugar, elevated cholesterol, body fluid balance, immune function, hormone metabolism, cognition, and many other metabolic processes. Adrenal stress is also associated with allergies, chronic stress, fatigue, anxiety, nervousness, muscle and joint pain, migraine headaches, sleep disturbances, and cardiovascular disease.

Electrolytes balance

As we have learned, many nutrients are depleted in the body while using cholesterol lowering drugs. Electrolytes are compounds made mostly of minerals (sodium, potassium, magnesium, calcium, etc.) that are dissolved within body fluids. Electrolytes are fundamental to life because all the cells of the muscles, nerves, and heart utilize them to maintain optimal function. Electrolyte imbalance can lead to serious health concerns such as dehydration, and as we discussed in chapter 6, electrolyte imbalance can ultimately lead to cardiovascular and neurological dysfunction.

Acid /Base Balance

All medical physiology texts recognize the pH balance of the body as one of the most important biochemical balances in human biochemistry. Since all enzymes, hormones, and cell receptors are dependent upon proper pH balance, all biochemical functions are under the influence of pH control.[64]

Body pH is very important because pH controls enzyme activity as well as the speed that electricity moves through the body. In fact, pH controls the speed of the entire body's biochemical reactions.

Dietary habits impact the body's pH levels. This in turn affects the activity of every metabolic function. For example, an acidic pH creates an environment ideal for the proliferation of yeast, viruses, bacteria, and cancer cells.

Other conditions associated with poor pH balance include bone loss, weak immunity, joint stiffness, achy muscles, chronic fatigue, cognitive decline, sleep disruption, acid reflux and various other digestive issues, cardiovascular disorders, atopic disorders, and allergies. Identifying and correcting pH imbalances can be a key component in restoring optimal health.

Cost of Care

Ask yourself this question: What is more cost effective—treating the *cause* of a problem or masking the sign or symptom? Let's look at the cost of treating the signs of altered physiology like high cholesterol with a statin drug instead of treating the cause.

The cost for a statin drug is between $2.50 to $5.00 for a single pill per day, which equals $75 or $150 a month or up to $1,800 a year.[65] Over 30 years, that's $54,000 for a drug that has proven to do nothing with regards to improving cardiovascular health. So people will be taking this drug at exorbitant cost for over 30 years while increasing their risk of death, *as the drug's only measurable effect is to make you more unhealthy!*

I believe a better investment for you is to find a practitioner with a different set of tools and make an investment in yourself for 6-12 months in order to transform your body. If you could transform your body in 6-12 months and undo what 20 years of unhealthy lifestyle habits have done, wouldn't you consider that a great return on your investment?

You can actually be a different person six months from now. I have seen it in my practice over and over. But it is going to take effort on your part because your functional medicine practitioner cannot eat for you, think for you, exercise for you, or take supplements for you. You must consider your relationship with your functional medicine practitioner as a partnership with both of you working toward the same goal of restoring your health and vitality.

Depending on the practitioner, there may be an average investment of anywhere from $800 to $2,200 in order to get you on the path of optimal health. But by making this investment and maintaining the newly adopted lifestyle, you can save an average of $51,800 over the next 30 years. This is possible by removing just *one needless drug* from your life as you make your body healthier. I cannot even imagine what the cost would be for prescription drugs over a 30-year time period for the average person using standard medical practices since the average American over the age of 65 is on five to six prescription medications.

In general, **Americans filled 11.5 prescriptions per person in 2008 while spending over $291 billion on these medications!**[66] The point is that there are over 2 million adverse reactions to prescription drugs per year, with 106,000 of those resulting in the patients' death.[67] If you invest in your health now and prevent illness, you will spend much less time, money, and heartache treating the lifestyle diseases that plague us today.

Imagine a life where there is no need to have a government bailout to cover the ever-increasing medical expenses of lifestyle diseases. Take charge of your own life. Be responsible for your own actions regarding the health of your body, because if you don't do it, no one will.

CHAPTER 9

REGENERATING YOUR BODY THROUGH LIFESTYLE CHANGES

If you couldn't tell by now, I am a big advocate for eating right, exercising, and thinking positively. And it has become obvious to me that the concept of eating right means different things to different people. I see every day what people really think living healthier looks like. It would be laughable if it weren't so sad. The average American's concept of eating better appears to me to be choosing diet soda over regular versions, or perhaps forgoing the Super-Sized version of their McDonald's meal, or including a tomato and some iceberg lettuce on their fast-food cheeseburger. However, I propose that getting our bodies to respond in dramatic fashion will require a complete paradigm shift in one's thinking, and perhaps not overnight but eventually a radical change in one's daily habits. If this sounds like something you are game for, this chapter is for you.

What I'm about to share with you are simply some basics. I cannot possibly go into detail about each of the following pointers, because that is beyond the scope of this book. You can, however, check the appendix for references if you want to dig deeper and study more on this topic.

Eating

Diet lays the foundation for health and vitality. As mentioned earlier, statistics from a Surgeon General's report indicate that 7 out of 10 leading causes of death in the U.S. are preventable and diet-related. In addressing chronic health concerns, it is imperative, of course, to focus on diet and lifestyle. The importance of wise food selection cannot be overstated, because each bite of food carries a message to each cell of the body. What message is being communicated depends on what kind of food is being ingested. Think of it this way: If your automobile requires diesel fuel, but you put in regular unleaded, it won't take very long at all for the automobile to respond negatively.

While our bodies are far more sophisticated and intricate than an automobile, the same concept applies nevertheless. The human body was not designed to function on chemical concoctions like bologna and chips and Velveeta cheese and Lucky Charms and other products in the endless assortment of man-made foods.

While the human body has the innate ability to make adjustments and function somewhat normally on junk food diets for long stretches of time in some cases, eventually system failures begin to occur. When system failure does occur, instead of changing the fuel that caused the breakdown in the first place, what most Americans tend to do is put duct tape on the check engine light, so to speak, by suppressing symptoms with one or more drugs.

What we eat matters. And simply shunning a cheeseburger in favor of salmon isn't enough. Having an understanding of good fats versus bad fats, knowing the difference between healthy fillers versus junk fillers, and knowing how to get an appropriate balance of favorable macronutrients (proteins, fats, fiber, and carbohydrates) is also important. Having an occasional indulgence of junk food won't kill you, but the majority of your diet should consist of foods that will sustain life, provide energy, and promote vitality.

Thus, the old adage still holds true: *You are what you eat.* Unfortunately, we tend to eat the wrong foods in our culture of convenience.

We have all heard that the golden rule in maintaining a healthy diet is to eat fruits, vegetables, and whole grains, and to minimize processed foods and sugar, saturated fat and trans-fats. While these guidelines are staples for a healthy food plan, they are still only part of the picture because even certain healthy foods cause the body to respond with higher levels of insulin. Insulin is a storage hormone that increases the fat storage and inhibits fat breakdown.

Also, even healthy foods when not in proper balance can cause other aspects of your internal regulation such as pH to either increase or decrease, causing adverse effects previously mentioned.

Finally, once the body has gone beyond its natural ability to buffer poor diet choices, what it needs in order to get back to normal cannot be fully obtained from diet alone in many cases. This is where food supplements, nutriceuticals (natural alternatives to drugs for clinical use), and exercise provide additional benefit.

So as we begin our list of hints for a healthier life, sometimes it is a good practice to consider what we are NOT eating that may be contributing to our health challenges. Rather than focusing on radical diets that sometimes border on starvation or focusing only on what one should *not* do, let's start by trying the simple approach of *adding* a few basic practices.

For example, if you feel you need a candy bar every day, replace the candy bar with a piece of fruit. Or if you drink soda, replace it with water. Even if you don't stop all the bad eating habits right away, just adding some nutrient-dense foods such as fruits, vegetables, whole grains, and lean meats can go a long way to make a difference.

Some quick Dietary Hints for a healthier life:
1. Eat food that is as close to its natural state as possible. (Avoid prepackaged foods, chips, cake, candy, etc...)

2. Look for 92% or leaner ground beef (grass fed), or use buffalo or deer meat.

3. Substitute almond, rice, or soymilk instead of cow's milk. If you must drink cow's milk, choose skim and/or one of the organic versions.

4. Use fruits for dessert instead of sugary cakes and cookies.

5. To retain the nutrients and fiber in vegetables, lightly steam them instead of boiling them.

6. Steam, bake, grill, or braise your foods instead of deep-frying.

7. Switch to healthier fats. Cut out lard, palm oil, and shortenings made with these oils. Instead, use healthy oils such as olive, canola, soybean, sunflower, safflower, sesame, peanut, and grape seed oil.

8. When baking, choose whole-wheat flour instead of all-purpose flour. The whiter the substance, the more processed it is, which means it contains much less nutrients and fiber.

9. Try this for one month: begin to read labels for everything you eat and avoid all foods with high fructose corn syrup in the ingredient list. You will probably be shocked to find out how many packaged foods contain

this ingredient, or some other type of sweetener like sucralose or aspartame. In an age where nearly all of our packaged foods are chemically-laden, it's important to know what you are putting in your body if you want to get healthy.

Exercise

Our bodies were designed to move. We have our own built-in intelligence system that prompts us to move and keep active. However, in general many people consciously or subconsciously suppress that intelligence. As Yogi Berra once said, "Whenever I get the urge to exercise I lay down until it passes."

The Industrial Revolution has played a major role as to why we are less active. People used to get their exercise from working in the fields and doing manual labor. Today, life is easier and more convenient as we have computers, cell phone, cars, remote controls, home delivery service, etc. These modern conveniences are *modern inconveniences* for our physical health. In today's society, we have to search for ways to replace the activity that technology has helped to remove. Thankfully, there are several simple ways to do this.

Quick and Easy Ways to Increase Your Activity Levels without Technically Going on an Exercise Program:

1. Park at the farthest spot in the parking lot when shopping or at work. (Not only will you walk a little bit

on the way to the door, it may save some dings to your car as well!)

2. Take the stairs instead of the elevator.

3. Get up and walk around while talking on the phone.

4. Yard work, while not necessarily fun for some of us, is something that needs to be done and is definitely a calorie burner.

5. Speed squats: Watch the clock and complete as many squats as possible in 5 minutes without stopping. Attempt to beat your previous number of squats the next time you try it.

6. While enjoying your favorite television program, throw in some push-ups, sit-ups, squats, or even light weight lifting during commercials.

7. Typing and desk work: While you are sitting, contract your gluteus muscles. Doing so you will burn calories and tone your buns! If you swap out your regular work chair for a Swiss ball, it will help your posture and abdominals, as the ball requires you to constantly tighten your stabilizing muscles in your torso to maintain your balance while sitting all day.

8. Take your loved one out dancing. Dancing is a great cardio workout that helps burn calories.

9. Playing with the kids or the dog: If you have children or a pet you can play with, this is the perfect opportunity to bond with them while burning some calories.

10. Swimming: If you don't want to swim laps or cannot swim, hang out in the shallow end and run around. Water resistance will do the rest.

The kind of exercise we do is an important consideration. While any movement is beneficial to counteract a sedentary lifestyle, there are ways to optimize your physical health.

You don't have to work out an hour every day, but if you simply put in 20 to 30 minutes daily or maybe 45 minutes on three predetermined days of the week, you will make more progressive gains. It is scientifically proven that small daily workouts will greatly improve your health even more effectively than a long workout a few days a week. Remember, even if you cannot currently do some exercise every day, some exercise is better than no exercise.

As you progress in your exercise habits, combine resistance training with cardio for the quickest and best results. Simply jogging or walking briskly is not enough if you wish to accelerate your progress. Think of Olympic athletes training multiple times per day to optimize their bodies so they can achieve a gold medal. Do you think merely jogging or brisk walking would help them reach their goal? Absolutely not.

That is why, in order for you to achieve your health goals, you must step up your efforts and progressively add some type of weight or resistance training as you intensify cardiovascular exercise to maximize your health.

Lack of physical activity is one of the major risk factors for cardiovascular diseases. Regular exercising makes your heart, like any other muscle, stronger. A stronger heart can pump more blood with less effort. Likewise, being physically fit helps with blood pressure. The way in which exercise can cause a reduction in blood pressure is not clearly understood, but all forms of exercise seem to be effective in reducing blood pressure. Aerobic exercise appears to have a slightly greater effect on blood pressure in hypertensive individuals than in individuals without hypertension.

The National Institutes of Health (NIH) report that even moderate exercise and physical activity can also improve the health of seniors who are frail, or who have diseases that accompany aging. Research also suggests that exercise and physical activity can help you maintain or partly restore your strength, balance, flexibility, and endurance.

Exercise even appears to have a positive influence on mental health. Dr. Daniel M. Landers stated, "We now have evidence to support the claim that exercise is related to positive mental health as indicated by relief in symptoms of depression and anxiety." [68]

Start with Baby Steps

It has been said that inch-by-inch anything's a cinch; but yard-by-yard it's hard. It would be wonderful if you could radically change your lifestyle overnight, but for many people that just isn't do-able, either because of the mental energy that it takes, or because of other life circumstances. So starting by making a few simple changes and then building on those as you go can pay huge dividends in the long run.

For example, if you are a chain-smoker, perhaps it would be asking too much to simply give up your cigarettes. Perhaps you've tried over and over and failed. But never having been a smoker myself, I have often wondered how successful a person might be at quitting if he or she would give up just one cigarette a day, for a total of seven less cigarettes per week. And then the following week do that again. Now you would be smoking 14 less cigarettes per week. And then do it again the following week, and the one after that, and the one after that. It seems to me that before long a smoker could quit fairly easily over time rather than doing it cold-turkey.

The same principle can be applied to other lifestyle issues. It has been said that the journey of a thousand miles begins with the first step. So if all this seems too overwhelming for you, don't sweat it. Simply begin with one or two changes that you know you can incorporate into your life fairly easily. You might begin, for instance, by just taking a pleasant walk around your neighborhood two or three times per week with the intention in mind of increasing your walking speed over time and/or increasing the number of days that you walk. And as for diet, you can begin by simply eliminating soda pop and build on that. Since there is no redeeming value in soda and it only leaches nutrients from the body, just making that one change could be huge for you. After you have that habit kicked, change something else in your diet, and continue over time until you have radically changed your lifestyle.

And finally, you could also easily incorporate the use of supplements in the diet. That alone could make a significant impact on your health. Supplements are the topic of our next chapter.

CHAPTER 10

SUPPLEMENTS THAT ENHANCE LIFESTYLE CHANGES

The term, *nutriceuticals*, is simply a fancy word for natural products used in the clinical setting as an alternative to drugs. These types of products, like their health food store counterparts, utilize various vitamins, minerals, and herbals in order to support various biochemical pathways and organ systems. Unlike health food store products, however, nutriceuticals are designed according to specific clinical applications and often are manufactured according to quality standards that exceed most over-the-counter brands, with a few exceptions.

Since Western diets are so nutritionally flawed and most likely cannot provide all of the nutrients we need even when we are eating better, nutritional supplements can fill in the gaps. Because of modern farming methods that often leave our grocery store produce devoid of the nutrients they would otherwise have under different growing conditions, even eating a whole foods diet cannot always provide the nutrition we need, unless you eat 100% organic, which is more expensive and less accessible.

In choosing nutritional supplements, quality is of utmost importance. *Consumer Reports* once reported that one-dollar out of every three spent on nutritional supplements is wasted because the quality control is so bad. Since the FDA does not regulate the supplement industry, standards of quality are left to each supplement manufacturer to decide for themselves. Many commercial supplement manufacturers use inexpensive ingredients of low biological value and questionable processing methods which result in products that are cheap to produce and more profitable to sell.

Furthermore, analyzing each batch of every product for all the various contaminants by a third-party laboratory is expensive and not required by the FDA, and is therefore not a practice implemented by most supplement companies.

There are not many who adhere to what I consider gold standard quality methods of manufacturing. While the list of quality control standards is too extensive to list here (not to mention boring), just keep in mind that the supplement industry is a business, just like the pharmaceutical industry, so the goal is to make money. While making money is not wrong, of course, there are those who do business with *only* profit in mind without much regard for the quality of the product and the biological value it can deliver. Thus, not everything on the shelves of your local pharmacy or health food store is going to deliver a significant impact on your health. That's why using doctor's line brands over the store-bought brands, by in large, is a better buy for your money. As has been aptly said, the most expensive supplement is the one that doesn't work. And, of course, the old saying remains true, *"You get what you pay for."*

In my practice I utilize one product line in particular that excels in quality and scientific innovation called, *Metagenics*. Although Metagenics is, for the reasons stated above, my primary pick of nutriceutical companies, it certainly isn't the only one that you could utilize for your nutritional needs. In the appendix section of this book, I will list some other companies that I recommend as well. For your convenience you will find all products available at www.drryanbentley.com, or talk to your Functional Medicine Practitioner to find out where you can get quality nutriceuticals.

Let's transition, then, and discuss some specific nutrients that can play a role in supporting cardiovascular health.

Co-Enzyme Q10 (CoQ10)

CoQ-10 is a fat-soluble vitamin-like substance present in every cell of the body and is a powerful antioxidant. It serves as a co-enzyme for several of the key enzymatic steps in the production of energy within the cells, especially the cells of the heart. CoQ-10 is naturally present in small amounts in a wide variety of foods, but is particularly high in organ meats such as heart, liver and kidney.

CoQ-10 is known to be highly concentrated in heart muscle cells due to the high-energy requirements of the heart. For the past several years the bulk of clinical work with CoQ-10 has focused on heart disease. Specifically, congestive heart failure from a wide variety of causes has been strongly correlated with low levels of CoQ-10.

More recently, research has focused on the importance of supplementing CoQ-10 in patients who are taking statin drugs for cholesterol, since statin drugs are known to cause deficiencies in CoQ-10 and thus predispose patients to heart failure as well as dysfunctions in other CoQ-10-dependant energy processes throughout the body, such as those of the brain and skeletal muscles. As stated in an earlier chapter, it has been proven that a person on statin drugs can go from a score of 84% in fatigue (meaning they are very fatigued) down to 16% just by giving CoQ-10. Supplementing CoQ-10 overcomes many of the side effects associated with statin drugs. [69,70,71,72,73,74,75,76,77,78,79,80]

When buying CoQ-10 supplements, be sure to choose either an oil-based gel cap or a liquid. Research has shown that CoQ-10 in the oil-base form is absorbed nearly 300% better compared to dry tablets or powder.

I personally use and recommend CoQ-10 100mg from *The Wellness Prescription*, as it is of highest quality and much less expensive than other companies of equal quality. Additionally, I sometimes recommend a liquid version of CoQ-10 from Metagenics called, *NanoCell-Q*.

Acetyl-L-Carnitine (ALC)

Carnitine is a nutrient that helps the body burn fat for energy. It is produced in the liver and kidneys and stored in the skeletal muscles, heart, brain and sperm.

Under normal circumstances, the body makes all the carnitine it needs. Some people, however, may be deficient in carnitine because their bodies cannot make enough of it or cannot transport it into tissues efficiently so that it can be used. Other conditions, such as angina or intermittent claudication (a clinical diagnosis for muscle pain), can also cause deficiencies of carnitine, as can some medications.

Acetyl-L-Carnitine (ALC), a derivative of carnitine, has been proposed as a treatment for many conditions because of its ability to reduce oxidative stress. ALC has the ability to cross the blood-brain barrier and get into blood-brain circulation where it acts as a powerful antioxidant, helping to prevent deterioration of neurons. Its supplementation has been shown to be protective for nerve cells in instances of cerebral ischemia (restriction of blood supply) in rats[81] and may be useful in treating peripheral nerve damage. ALC has also been shown to improve insulin response, which has implications in the prevention of heart disease and oxidation, and has proven to have a positive effect on various muscle diseases as well as heart conditions. [82]

As it pertains specifically to heart disease, ALC helps the body to produce energy by "feeding" the mitochondria, the little energy-producing factories within our cells. In fact, without carnitine, a specific kind of fat called *long chain fatty acids* would not be able to enter the mitochondrial membrane. Long chain fats need carnitine to transport them through the membrane so they can be metabolized. Studies have shown that muscles generate up to 70% of their energy through the burning of fats; and since the heart is the hardest working muscle in the human body, it makes sense that ALC can be so beneficial for it. There have, in fact, been many well-designed studies which suggest that ALC may be beneficial for a number of different cardiac problems.

It is without question that ALC benefits the mitochondrial processing of fats, which fuels cellular energy. However, greater energy also means greater electron flow which may cause increased oxidation. The solution in laboratory research is to co-supplement ALC with other antioxidants. This helps enhance the reversal of age-related degeneration while further decreasing oxidation. That's one reason why I do not prescribe ALC by itself, but prescribe a product with a combination of antioxidants. Powerful nutrients to consider adding to

ALC in a product would be N-acetylcysteine (NAC), grape seed extract, and specific B-vitamins that support homocysteine metabolism, providing another benefit to the cardiovascular system.

Fish Oil and Heart Disease

One group of dietary nutrients enjoying much publicity as it pertains to inflammation and other health benefits are the omega fatty acids. Don't be confused by the wording here. Omega-3 fatty acids are anti-inflammatory because they displace arachidonic acid (bad fat) and other inflammatory chemicals, while omega-6 acids can actually drive inflammation. Proper balance is the key. We already get too many omega-6 fatty acids in the diet, but we have dangerously low levels of omega-3s. Dietary sources of omega-6 fatty acids would be meat, eggs, dairy and certain vegetable oils such as the ones found in corn, safflower, and sunflower. Foods rich in omega-3's include oily fish such as herring, sardines, tuna, mackerel, salmon (preferably wild) and oils made from these fish; also hemp, flax, pumpkin seeds, soy, canola oil and walnuts and their oils. Other sources for small amounts of omega3 fatty acids are non-starchy vegetables such as dark leafy salad greens, spinach, kale, tomatoes, broccoli, cauliflower, collard greens and onions.

A healthy ratio of these two fatty acids would be 1:1, or at most 5:1 in favor of the omega-6s. However, the ratio in the typical American diet is more than 20:1 in favor of the omega-6 fats. This unfavorable ratio shifts the body toward a chronic inflammatory state which can set the stage for many chronic conditions.

Scientists were first alerted to the importance of omega-3 fatty acids in the 1970s when Danish researchers discovered that Greenland Eskimos suffered almost no heart disease, cancer, or arthritis in spite of eating diets very high in animal protein and fat. It was later discovered that most of the foods they were eating were rich in omega-3 fatty acids and of two omega-3 fatty acids in particular - Eicosapentaenoic acid (EPA), and Docosahexaenoic acid (DHA).

While altering the diet is paramount, we can also achieve more therapeutic levels of omega-3 fatty acids by supplementing our diet with purified fish oil. Some research suggests taking as much as 4 grams per day of fish oil for cardiovascular disease, and 8 grams for neurological diseases. It is nearly impossible to get that much omega-3 fatty acids from diet alone, so supplementing becomes necessary as a way to manage these types of problems.

The benefits of fish oil supplementation enjoy one of the broadest and largest bodies of evidence of any therapeutic substance, pharmaceutical or natural. In fact, a report from the Journal of the American Medical Association demonstrated that fish oil supplementation has been shown to reduce the risk of heart attacks by up to 90% and increase IQ scores by up to 13%.

EPA and DHA, categorized as omega-3 fatty acids, bring about their health properties by benefiting the body in several ways: First, omega-3 fatty acids compete with other chemicals in the body that can cause an inflammatory response, such as arachidonic acid, which is present in dairy and meat products, is also produced by the body during inflammatory responses. As mentioned before, chronic inflammation has been implicated in heart disease and other degenerative conditions. EPA and DHA help to drive out arachidonic acid and thus help to cool inflammation, which may help to relieve minor pain as well.

EPA and DHA also help to thin the blood, thus helping to support healthy blood pressure. They likewise help prevent blood platelets from becoming sticky, thereby helping to prevent the accumulation of plaque in arteries that can lead to strokes and heart attacks. [83,84,85,86,87,88,89,90,91,92,93,94,95,96,97]

One issue to be aware of in choosing fish oil supplements, however, is contamination. The FDA has stated that a 132-pound woman eating two cans of albacore tuna per week would exceed safe dose ranges of mercury by 3 times! Unfortunately, one of God's most healthful foods for human consumption has been contaminated by various toxins and heavy metals. For this reason, eating fatty fish must be limited, and this emphasizes the importance of supplementing with fish oil capsules. However, fish oil supplementation does not necessarily solve the problem, as fish oil supplements can be *more* toxic than eating fish if manufacturers do not take the necessary steps to remove all the contaminants through an ultra-purification molecular distillation process and then follow up on that process by paying a laboratory to perform a third-party analysis verifying that the oils are contaminant free.

Unfortunately, quality fish oil supplements are rare. Fish oil capsules often contain high amounts of mercury and other various heavy metal contaminants. As mentioned earlier, there are no specific regulations regarding purity in nutritional supplements. It is important to obtain purity-certified fish oil supplements that are verified by third-party laboratory analysis. Likewise, third-party laboratory analysis proving that the amount of EPA and DHA claimed on the label is actually represented in the product is also an important determinant of quality.

While I typically do not recommend over-the-counter products to patients, there is one over-the-counter fish oil line of supplements that is doing an excellent job along these lines called, *Nordic Naturals*. Metagenics also has an excellent line of fish oil products.

Alpha Lipoic Acid

Alpha lipoic acid is a vitamin-like antioxidant that is produced by the body and found in certain foods. One of its primary roles is converting glucose into energy, but this unique antioxidant is involved in many secondary functions. Its antioxidant properties help support the cardiovascular system. Most antioxidants work in either water (vitamin C, grape seed extract, green tea) or fatty tissues (carotenoids, CoQ10, vitamin E). Alpha lipoic acid functions in both water and fat. This allows it access to every sort of tissue in the body.

Another important trait of lipoic acid is its ability to help regenerate other antioxidants. It therefore allows the antioxidants that are eaten and taken as supplements to work harder and longer in the body, further decreasing the possibility of the oxidation of cholesterol.

The most common conditions for which lipoic acid has been used are mostly related to diabetes and liver disorders. Lipoic acid is known to enhance blood sugar control by improving insulin sensitivity. Thanks to its anti-inflammatory effects, it's also been shown to help with a painful diabetic complication called neuropathy (nerve damage). In addition, its powerful antioxidant properties seem to provide a protective effect in a number of liver diseases.

Recently scientists have been experimenting with the use of alpha lipoic acid in a broader range of conditions.

Oregon State University conducted a trial to evaluate the potential of alpha lipoic acid in lowering high triglycerides (a risk factor for heart disease). The results revealed a drop in triglycerides of up to 60%.

Here's a summary of a few other studies on the cardiovascular benefits of alpha lipoic acid:

- A study found that lipoic acid could help lower cholesterol and lipid peroxidation, a process that contributes to plaque forming in arteries.[98]
- Another 2008 study published in the *British Journal of Pharmacology* revealed an improvement in endothelial function in rats fed alpha lipoic acid. The endothelium is a thin layer of cells in our blood vessels that helps blood to circulate properly and therefore assists in promoting healthy blood pressure. The authors of this study proclaimed that, "The favorable antioxidant, anti-inflammatory, metabolic and endothelial effects of lipoic acid shown in rodents, in this and other recently published studies, warrant further assessment of its potential role for prevention and treatment of cardiovascular diseases."[99]
- The prior study lends even more support from a review article, which highlights alpha lipoic acid as a potential superstar in the management of hypertension (high blood pressure).[100]

I recommend for my patients a product called Meta Lipoate 300.

Medical Foods for Detoxification

In our day and age, toxins in the environment and in our food supply are so abundant that the body cannot keep up with detoxifying them and eliminating them properly. However, there are wonderful nutritional products that can enhance the body's clearance of toxins and thus keep oxidation and fat accumulation under control.

A ground-breaking formula for this purpose created by Metagenics is the product, *UltraClear Plus*. This medical food helps support and balance the liver detoxification pathways for more efficient clearance of toxins, even when the toxin load and accumulation is significant. Boasting dozens of micro-nutrients such as B6, B12, activated folate and magnesium, to name just a few, *UltraClear Plus* is a highly-researched product designed to fuel the various pathways in the liver that are responsible for the process of biotransformation, or the conversion of toxins into harmless substances that can easily be eliminated by the body.

A follow-up product introduced years after the first product was developed is called *UltraClear Plus PH*. This formula has higher amounts of potassium, magnesium, and calcium, along with the addition of sesamin for battling acidity and encouraging alkalinity. Sesamin is an especially important addition to the mix because it plays a powerful downstream role in enhancing toxin conversion and elimination.

As a complement to the UltraClear line of products, the nutriceutical product, *AdvaClear* by Metagenics, provides plant substances such as ellagic acid from pomegranate, Silymarin from Milk Thistle, and Watercress, among others, that provide a re-balancing effect to the liver detoxification pathways that is unique even compared to *UltraClear Plus*. Also, the addition of such substances as N-acetylcysteine, which helps the body make glutathione (a powerful anti-oxidant) and an important compound for the body's de-toxification abilities, makes AdvaClear a gold standard for aiding the body's ability to clear foreign molecules and a powerful antioxidant formula as well.

Medical Food to Quench Inflammation

Remember that the total cholesterol level is not nearly as important as inflammation and oxidation. Thus, there is a product I have used for years in my nutritional armament that has the ability to kill two birds with one stone. The ground-breaking medical food by Metagenics, called *UltraInflamX Plus 360,* is a product designed to support the body's fight against inflammation. The fringe benefit of the ingredients included to fight inflammation is the huge antioxidant value it has.

There is a measurement of antioxidant potential in food and supplements called the ORAC value. ORAC stands for *Oxygen Radical Absorbance Capacity.* The higher the ORAC value, the better. Some antioxidant-rich foods like cranberries have a high ORAC value of 9,000 or so. The UltraInflamX product boasts an incredible ORAC value of over 17,000 per serving! So not only can UltraInflamX help address chronic systemic inflammation, but it can also provide an enormous antioxidant boost. Substances in UltraInflamX, such as turmeric, for example, are nature's natural anti-inflammatory agents; but turmeric also happens to have some of the most powerful antioxidant potential of any substance on the planet.

UltraInflamX Plus 360 also features rosemary, another one of the world's most powerful antioxidant substances. Combining this with extracts of hops and olive leaf, along with dozens of micronutrients, this product is like napalm to free radicals and inflammation. As a side benefit, the rosemary and hops extracts also have some mild blood thinning properties, further benefitting cardiovascular function by preventing blood platelets from becoming sticky. While expensive, I know of no better product to cool the body's inflammation while also providing antioxidant value, which according to the research is the true cause of heart disease.

Getting Started

If you had to narrow down your nutritional regime to just a couple of things because of cost, it would most certainly be a high-potency fish oil product, and probably CoQ10. Of course, having a good multi-vitamin on board would also help to provide foundational support. So by adding just a few basic products, along with healthy lifestyle changes, you can go a long way in impacting your health for years to come.

Please refer to the index for other nutritional lines to consider.

CHAPTER 11

SUMMARY AND CONCLUSION

We've covered a lot of ground, so let's briefly review some primary points as we bring our discussion to a close.

- Cholesterol is important for many different functions of the body.
- The lowest levels of cholesterol throughout the world belong to countries where heart disease tends to be the highest, whereas rates of heart disease are lower in some other countries where cholesterol tends to be higher.
- Heart disease is caused by oxidation leading to inflammation.
- Oxidized cholesterol in an inflamed system is the real culprit in heart disease, not the levels of cholesterol.
- Statin drugs are known carcinogens (cancer-causing

agents), and are also associated with increased risk of congestive heart failure, as well as neurological disorders like Parkinson's and memory loss.
- Research shows that statin drugs do not decrease one's risk of having a heart attack.
- Research clearly shows that statin drugs and lower levels of cholesterol lead to early death.
- Statin drugs continue to be marketed in spite of their safety and efficacy record. This can be linked in all likelihood to the financial interests of the FDA review boards and the executives of statin manufacturers.
- Diet and lifestyle are the major culprits in heart disease and most chronic diseases.
- Diet and lifestyle can be used successfully and without side effects to manage and even reverse heart disease in many cases.
- Dietary changes should be focused on eating more fresh, whole foods in their most natural forms.
- Exercise is a major component of cardiovascular rehabilitation and overall health. Even taking a brisk walk around your neighborhood 3 or 4 times per week can reap huge health dividends.
- Lifestyle modification can begin slowly with just a few changes made at first and progressing over time to more significant changes. Beginning by just adding a walk in the evening a few days per week and eliminating soda pop can be important first steps on your health journey.
- Specific and broad-spectrum supplementation can support various biochemical pathways and organ

systems that may have grown weak over time, thus helping to accelerate your progress and restore your health.
- The most basic supplements to consider for cardiovascular health are fish oil, CoQ10, and a broad-spectrum multiple vitamin formula. Other more specific nutrients can be added if necessary, according to the suggestions in the previous chapter when you are properly evaluated.

R. Buckminster Fuller once said, "You never change things by fighting the existing reality. To change something, build a new model that makes the existing model obsolete." How is that done? You do that by telling a better story, one based upon truth vividly demonstrated. That's what I hope this book has provided for you – a better and more truthful story about cholesterol and the true causes of heart disease, and better methods of fixing those causes.

I sincerely hope this discussion bears fruit in your life and gets you or a loved one started on the road to recovery and optimal health.

REFERENCE LIST

1. *Principles of Anatomy and Physiology.* 12th edition. Tortora and Derrickson. p. 48-49, 991
2. ibid
3. ibid
4. ibid
5. ibid
6. ibid
7. ibid
8. ibid
9. Davis, Adelle. *Let's Get Well.* Harcourt Brace Jovonovich, INC. 1965. p. 57.
10. Ross, R. "Atherosclerosis – An Inflammatory Disease." *New England Journal of Medicine.* 1999. 340: pp.115-26
11. *American Journal of Cardiology.*1 August 2003:92(3)
12. Krauss, Dr. Ronald M., *Director of Atherosclerosis Research at the Oakland Research Institute*
13. *The American Journal of Epidemiology. 1 Sept. 2004: 160(5):408-420*
14. Eaton, S. Boyd M.D. and Konner, M. Ph.D. "Paleolithic Nutrition: A Consideration of its Nature and Current Implications." *New England Journal of Medicine.* 1985. 312, 283-289

15. *American Journal of Preventive Medicine.* Nov-Dec 1991:7(6):406-9.
16. *Journal of the American College of Cardiology.* 2002. 39:1567-1573
17. *Journal of the Royal Society of Medicine.* Oct. 2004. Vol. 97, No. 10. pp. 461-464
18. ibid
19. ibid
20. ibid
21. *The Lancet.* Vol. 358, No. 9279. 4 Aug. 2001. pp. 351-355
22. ibid
23. Atrens, D.M. "The Questionable Wisdom of Low-fat and Cholesterol Reduction." *Social Science & Medicine. Aug.1994. 39(3):433-47.*
24. *The Lancet. 2007. 369: 268-9*
25. ibid
26. McNamara, Dr. Ladd, "The Cholesterol Conspiracy" pp.34-41
27. *Journal of BioFactors,* 2005. 25(1-4): pp. 147-152.
28. ibid
29. Ross, R. "Atherosclerosis – an inflammatory disease." *New England Journal of Medicine.* 1999. 340:115-26
30. *Nutrition, Metabolism and Cardiovascular Diseases,* Vol. 15, No. 1. pp. 36-41
31. *Prostaglandins, Leukotrienes and Essential Fatty Acids.* Vol. 71, No. 4, Oct. 2004. pp. 263-269
32. *New England Journal of Medicine.* 18 March 2010.
33. ibid
34. Hayward, Dr. Rodney A., *Professor of internal medicine at the University of Michigan Medical School.*
35. ibid
36. *Journal of the American Medical Association.* 3 Jan. 1996. 275(1):55-60
37. *Journal of the Royal Society of Medicine.* Oct. 2004. Vol. 97, No. 10, pp. 461-464
38. Hadler, Dr. Nortin M., *Professor of Medicine at the University of North Carolina at Chapel Hill*

39. *Business Week. 17* Jan. 2008
40. *Principles of Anatomy and Physiology 12th Edition. Tortora and Derrickson.* p. 773
41. *Healthy Day.* 22 Aug. 2008
42. *NaturalNews.* Link to full Canadian Medical Association Journal article: <http://www.cmaj.ca/cgi/content/full/179>
43. *Journal of the American Medical Association.* 3 Jan. 1996. 275(1):55-60
44. *Journal of Sexual Medicine* 2010;7:1547–1556.
45. *Circulation.* March 2006. 113 (12): 1553–5
46. *European Journal of Clinical Investigation.* 1998. 28: 235-42.
47. *Drug-Induced Nutrient Depletion Handbook.* 2nd ed.
48. ibid
49. ibid
50. ibid
51. ibid
52. ibid
53. ibid
54. ibid
55. ibid
56. ibid
57. ibid
58. ibid
59. ibid
60. ibid
61. *Time,* 23 Feb. 23. pp. 38-46
62. *U.S. Department of Agriculture (USDA).* Data: Sugar Consumption in 1999
63. *Nutrition & Metabolism.* 2008. 5:2
64. *Journal of Endocrinology* 1990:128: 2476-2488
65. "The Statin Drugs Prescription and Price Trends," *Consumer Report* November 2004 to October 2005. Reported Jan. 2006

66. *most-medicated-states-lifestyle-health-prescription-drugs* http://www.forbes.com/2009/08/17/
67. *Journal of the American Medical Association (JAMA).* Vol. 284, No. 4. 26 July 2000.
68. *The Influence of Exercise on Mental Health*, Daniel M. Landers, ARIZONA STATE UNIVERSITY
69. *Journal of Clinical Pharmacology.* 1993. 33, 3, pp. 226-229.
70. *Proceedings of the National Academy of Science.* Vol. 87. pp. 8931-8934.
71. *Proceedings of the National Academy of Science, U.S.A.* 1985. Vol. 82(3), pp. 901-904.
72. *Drugs Experimental Clinical Research.* 1984 X (7) 497-502.
73. ibid
74. *Drugs Experimental Clinical Research.* 1985 Vol. 11(8), pp. 581-593.
75. "Effective and safe therapy with coenzyme Q10 for cardiomyopathy." In: *Klin. Wochenschr.* 1988 66:583-593.
76. *The American Journal of Cardiology.* 1989. Vol. 65, pp 521 - 523.
77. Int. J. Tissue React. 1990. Vol. 12 (3) pp. 155-162.
78. "Coenzyme Q10 treatment of heart failure in the elderly: Preliminary results." *Biomedical and Clinical Aspects of Coenzyme Q*, 1991. Vol. pp. 473-480.
79. "Effect of coenzyme Q10 on left ventricular function in patients with dilative cardiomyopathy." *Current. Therapeutic Research.* 1991. 49:878-886.
80. *The Molecular Aspects of Medicine*, 1994 Vol. 15 pp. S165-S175.
81. *Clinical and Experimental Pharmacology & Physiology* August 2006. 33 (8): 725–33.
82. *Hypertension* Sept. 2009 54 (3): 567–74.
83. "n-3 polyunsaturated fatty acids and the cardiovascular system." *Current Opinion Lipidol.* 2000. 11(1):57-63.
84. *Atherosclerosis.* Nov. 2006. 189(1):19-30.

85. *American Journal of Cardiology.* 2007. 99(6A):S35-43.
86. *Journal of Membrane Biology.* 2006. 206:155-63.
87. *Journal of the American College of Cardiology.* 2005. 45:1723-8.
88. "Evaluation of the antihyperlipidemic properties of dietary supplements." *Pharmacotherapy.* 2001. 21(4):481-487.
89. *Circulation.* 2001. 103:623-625.
90. *American Journal of Clinical Nutrition.* 2001. 74(4):464-473.
91. "Fish consumption, omega 3 fatty acids and cardiovascular disease. The science and the clinical trials." *Nutrition Health.* 2009. 20(1):11-20. Review.
92. *American Journal of Clinical Nutrition.* 2005. 81:416-20.
93. *Current Opinion Lipidol.* Feb. 2009. 20(1):30-8.
94. "Omega 3 fatty acids for prevention and treatment of cardiovascular disease." *Cochrane Database Syst Rev.* 2004:CD003177
95. *Journal of the American Medical Association (JAMA).* 2001. 285(3):304-312.
96. *Journal Nutrition.* March 2009. 139(3):495-501.
97. *Nutrition Health.* 2009:20(1):41-9. Review.
98. *Biomed Pharmacother.* Dec. 2008 62(10):716-22. Epub 4 Jan 2007.
99. "Lipoic acid supplementation and endothelial function." *British Journal of Pharmacology.* Apr. 2008, 53(8):1587-8. Epub 17 March 2008.
100. *A Systematic Review Clinical and Experimental Hypertension,* Issue 7, October 2008, pp. 628 – 639.

APPENDIX

How to locate Functional Medicine Practitioners:
1. www.wellnessprescription.net
2. www.functionalmedicine.org/findfmphysician/index.asp

Sources for more lifestyle changes:
1. The Vessel "Choosing the lifestyle that bears good fruit" by Dr. Ryan E. Bentley to be released 2011
2. The Eat-Clean Diet by Tosca Reno
3. http://dunamisfitness.wordpress.com

Where to purchase quality nutritional supplements (nutriceuticals):
1. www.drryanbentley.com
2. www.twp.meta-ehealth.com

Supplement Companies with good quality:

Upper-Level Professional Nutriceutical Lines
1. Metagenics: 800-692-9400
   www.metagenics.com
2. Nutri-Dyn: 800-444-9998
   www.nut-dyn.com
3. Anabolic Labs: 800-344-4592
   www.anaboliclabs.com

4. Biotics Research Corporation: 800-231-5777
   www.bioticsresearch.com
5. Xymogen: 800-647-6100
   www.xymogen.com
6. Integrative Theraputics: 800-332-2351
   www.integrativeinc.com
7. Standard Process: 800-558-8740
   www.standardprocess.com

Over-the-Counter Lines
1. Nordic Naturals: 800-662-2544
   www.nordicnaturals.com
2. Carlsons: 888-234-5656
   www.carlsonlabs.com
3. Enzymatic Therapy: 800-783-2286
   www.enzymatictherapy.com
4. Ethical Nutrients: 800-6668-8743
   www.ethicalnutrients.com

Claim <u>Your FREE Gifts</u> That Accompany *Sex, Lies, & Cholesterol*
Go To: www.DrRyanBentley.com

Visit DrRyanBentley.com and claim your free gifts, download additional resources, and learn more about Dr. Bentley's crusade to save our nation's health.

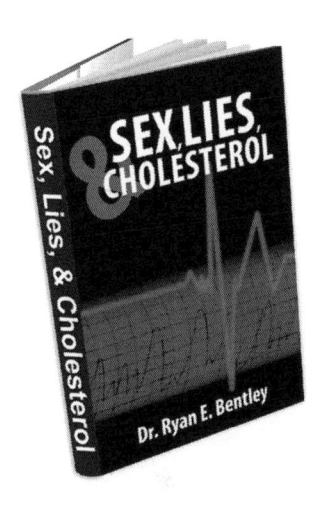

SEX, LIES, and CHOLESTEROL

Dr. Ryan Bentley, one of the Nation's foremost experts on Wellness, tackles widely held medical beliefs in this mind-altering book, revealing what many believe to be fact as pure myth. Learn more at DrRyanBentley.com

www.DrRyanBentley.com

Manufactured By: RR Donnelley
                   Momence, IL USA
                   January, 2011